Communications in Asteroseismology

Volume 161
Compendium, January to June, 2010

Communications in Asteroseismology
Editor-in-Chief: **Michel Breger**, michel.breger@univie.ac.at
Editorial Assistant: **Isolde Müller**, isolde.mueller@univie.ac.at
Layout & Production Manager: **Isolde Müller**, isolde.mueller@univie.ac.at
Language Editor: **Natalie Sas**, natalie.sas@ster.kuleuven.be

CoAst Editorial and Production Office
Türkenschanzstraße 17, A - 1180 Wien, Austria
http://www.oeaw.ac.at/CoAst/
Comm.Astro@univie.ac.at

Cover Illustration

Echelle diagram of frequencies identified by a Fourier iterative sine-wave fitting algorithm (left) and a Bayesian Markov Chain Monte Carlo algorithm (right) for a simulated time series with problematic aliasing. Input frequencies (blue), their aliases (green) and identified frequencies (red) are shown. (Illustration kindly provided by T. White. For more information see the paper by T. White et al., page 39)

British Library Cataloguing in Publication data.
A Catalogue record for this book is available from the British Library.

ISBN 978-3-7001-6915-4
ISSN 1021-2043

Austrian Academy of Sciences Press
A-1011 Wien, Postfach 471, Postgasse 7/4
Tel. +43-1-515 81/DW 3402-3406, +43-1-512 9050
Fax +43-1-515 81/DW 3400
http://verlag.oeaw.ac.at, e-mail: verlag@oeaw.ac.at

Comm. in Asteroseismology
Volume 161, Compendium 2010

Introductory Remarks

Starting with CoAst 160, we have changed our procedure in order to achieve immediate publishing. As soon as a paper is accepted by the referee and the editorial office, it is published electronically through ADS and can also be downloaded from our web site. This also means that the author will be able to cite his/her paper very quickly (and hopefully obtain a raise in salary as in some countries the number of publications count.) All the new accepted and electronically published papers will be collected periodically and distributed in a printed volume. However, the date of publication is the date of the electronic publication, not the printing.

We hope that this will lead to much faster publication. There will no longer be a deadline for our papers; just submit your paper when it is finished. By the way, we are not the only journal changing to this system.

Michel Breger
Editor-in-Chief

Contents

Introductory Remarks
by Michel Breger, Editor iii

Scientific Papers

Scaled oscillation frequencies and échelle diagrams as a tool for comparative asteroseismology
by T. R. Bedding and H. Kjeldsen 3

Photometric observations and frequency analysis of the δ Scuti star IP UMa
by D. Sinachopoulos, P. Gavras, Chr. Ducourant 17

Photometric and Spectroscopic Study of the δ Scuti Stars FH Cam, CU CVn and CC Lyn
by G. J. Conidis, K. D. Gazeas, C. C. Capobianco, W. Ogloza 23

A comparison of Bayesian and Fourier methods for frequency determination in asteroseismology
by T. R. White, B. J. Brewer, T. R. Bedding, D. Stello and H. Kjeldsen 39

A new eclipsing binary system with a pulsating component detected by CoRoT
by K. Sokolovsky, C. Maceroni, M. Hareter, C. Damiani, L. Balaguer-Núñez, and I. Ribas 55

Scientific
Papers

Comm. in Asteroseismology
Volume 161, January 2010

Scaled oscillation frequencies and échelle diagrams as a tool for comparative asteroseismology

Timothy R. Bedding[1] and Hans Kjeldsen[2]

[1] Sydney Institute for Astronomy (SIfA), School of Physics, University of Sydney, NSW 2006, Australia
[2] Danish AsteroSeismology Centre (DASC), Department of Physics and Astronomy, Aarhus University, DK-8000 Aarhus C, Denmark

Abstract

We describe a method for comparing the frequency spectra of oscillating stars. We focus on solar-like oscillations, in which mode frequencies generally follow a regular pattern. On the basis that oscillation frequencies of similar stars scale homologously, we show how to display two stars on a single échelle diagram. The result can be used to infer the ratio of their mean densities very precisely, without reference to theoretical models. In addition, data from the star with the better signal-to-noise ratio can be used to confirm weaker modes and reject sidelobes in data from the second star. Finally, we show that scaled échelle diagrams provide a solution to the problem of ridge identification in F-type stars, such as those observed by the CoRoT space mission.

Accepted: January 28, 2010

Individual Objects: η Boo, α Cen A, α Cen B, β Hyi, τ Cet, HD 49385, HD 49933, HD 181420, HD 181906

1. Introduction

This paper discusses how to compare frequency spectra of oscillating stars. We focus on solar-like p-mode oscillations, in which mode frequencies generally follow a regular pattern. This makes it useful to characterize them by a handful of frequency separations: the so-called large separation $\Delta\nu$ between consecutive overtones of a given angular degree l, and the small separations between adjacent modes of different degree. These frequency separations have the advantage of being closely related to physical properties of the stellar interior (see Section 2.). Measuring them and their variations with frequency and comparing with theoretical models is a major focus of asteroseismology (see, for example, reviews by Brown & Gililand 1994 and Christensen-Dalsgaard 2004).

There is somewhat less focus on absolute frequencies, partly because stellar models do not properly model the near-surface layers (Christensen-Dalsgaard et al. 1988; Dziembowski et al. 1988; Rosenthal et al. 1999; Li et al. 2002). This makes it difficult to compare individual observed frequencies with models, although Kjeldsen et al. (2008) have proposed an empirical correction that appears promising, at least for stars reasonably close in effective temperature to the Sun.

Here, we wish to compare observations of one star with observations of another, and so difficulties with models are not relevant. We are motivated by the expectation from homology that if two stars are sufficiently similar, their oscillation frequencies will be in the same ratio as the square roots of their mean densities:

$$\frac{\nu_1}{\nu_2} = \sqrt{\frac{\bar{\rho}_1}{\bar{\rho}_2}}. \tag{1}$$

Here, we are comparing modes in the two stars with the same radial order (n) and angular degree (l). Even if the two stars are not similar, we might still expect Equation 1 to provide a useful approximation. Of course, it also follows that the large separation scales in the same way:

$$\frac{\Delta\nu_1}{\Delta\nu_2} = \sqrt{\frac{\bar{\rho}_1}{\bar{\rho}_2}}. \tag{2}$$

We now present some examples and applications, using échelle diagrams to visualize the comparisons between stars.

2. Échelle diagrams and the asymptotic relation

The échelle diagram, first introduced by Grec et al. (1983) for global helioseismology, is nowadays used extensively in asteroseismology as a valuable way of displaying oscillation frequencies. It involves dividing the spectrum into segments of length $\Delta\nu$ and stacking them one above the other so that modes with a given degree align vertically in ridges. Any departures from regularity, such as variations in the large separation with frequency, are clearly visible as curvature in the échelle diagram, and variations in the small separations appear as a convergence or divergence of the corresponding ridges.

We conventionally define three observable small frequency separations: $\delta\nu_{02}$ is the spacing between $l = 0$ and $l = 2$; $\delta\nu_{13}$ is the spacing between $l = 1$ and $l = 3$; and $\delta\nu_{01}$ is the amount by which $l = 1$ is offset from the midpoint of the $l = 0$ modes on either side. In practice, the large and small separations are observed to vary with frequency.

The regularity in solar-like oscillation spectra allows us to write the mode frequencies in terms of the large and small separations, as follows:

$$\nu_{n,l} = \Delta\nu(n + \tfrac{1}{2}l + \epsilon) - d_l, \qquad (3)$$

where ϵ is a dimensionless offset. The small separation d_l is zero for $l = 0$ (radial modes), and equals $\delta\nu_{01}$ for $l = 1$, $\delta\nu_{02}$ for $l = 2$ and $(\delta\nu_{01}+\delta\nu_{13})$ for $l = 3$. Bedding et al. (2010b) have suggested that this last separation should be designated $\delta\nu_{03}$.

Equation 3 describes the oscillation frequencies from an observational perspective. A theoretical asymptotic expression (Tassoul 1980; Gough 1986, 2003) relates $\Delta\nu$, d_l and ϵ to integrals of the sound speed. In particular, $\Delta\nu$ measures quite accurately the sound travel time across the diameter of the star, while the small separations are sensitive to the structure of the core and ϵ is sensitive to the surface layers.

When making an échelle diagram, it is common to plot ν versus (ν mod $\Delta\nu$), in which case each order slopes upwards slightly. However, for grayscale images it can be preferable to keep the orders horizontal. We have done that in this paper, and so the value given on the vertical axis is actually the frequency at the middle of the order.

3. Scaled échelle diagrams and their applications

The ridges in an échelle diagram will only appear vertical if we use the correct value of $\Delta\nu$. For this reason, it does not generally make sense to plot two stars on the same échelle diagram. However, if the frequencies of the second star have been scaled by multiplying them all by the ratio of the large separations, we are led by Equation 1 to expect that its ridges can be made to coincide with those of the first. The scaling factor can be fine-tuned to optimize the alignment in two different ways:

1. by matching the slopes for the two stars (making both vertical), which means matching $\Delta\nu$, or

2. by overlaying the ridges as closely as possible, although this may mean they have different slopes, which means forcing them to have the same value of ϵ.

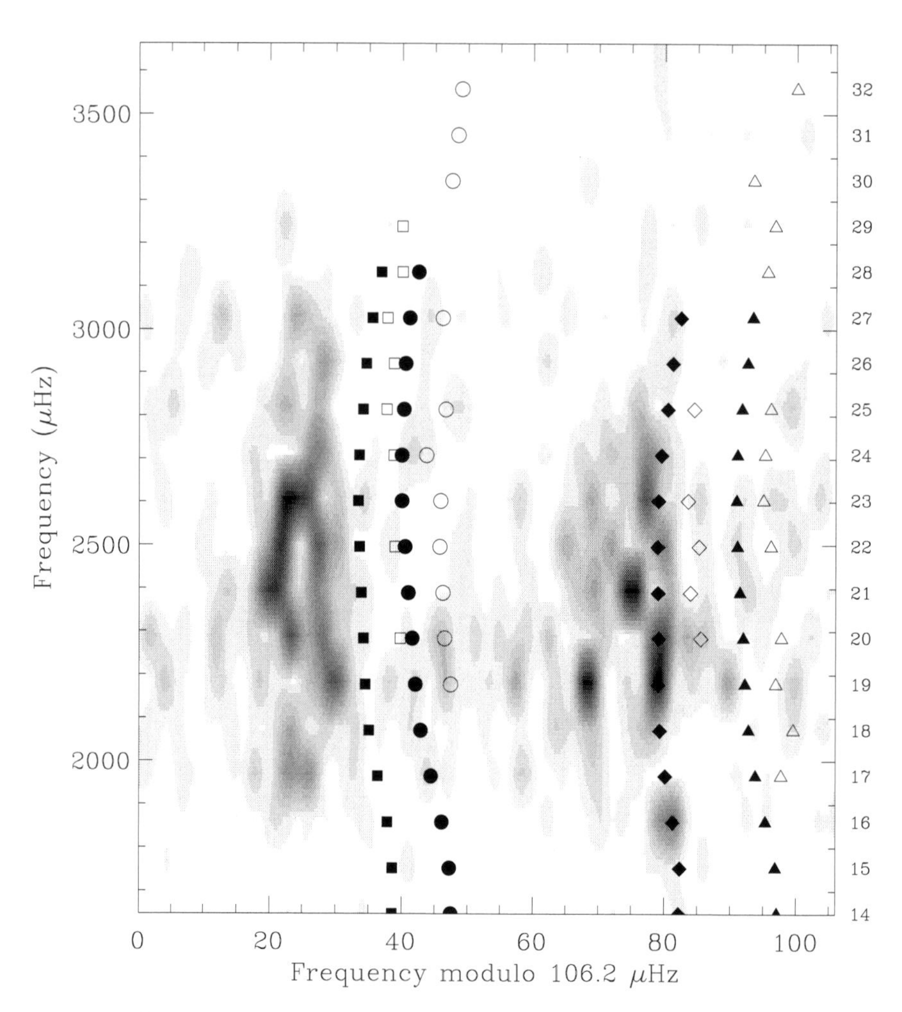

Figure 1: Scaled échelle diagram comparing three main-sequence stars. The greyscale is the power spectrum of α Cen A, filled symbols are oscillation frequencies for the Sun after multiplying by 0.7816, and open symbols are frequencies for α Cen B after multiplying by 0.6555 (see text for references). The scaling factors were fine-tuned to make the ridges for all three stars parallel (Method 1). Symbol shapes indicate mode degree: $l = 0$ (circles), $l = 1$ (triangles), $l = 2$ (squares) and $l = 3$ (diamonds).

Method 1 is shown in Figure 1 for three main-sequence stars: α Cen A ($\Delta\nu \approx 106\,\mu$Hz), the Sun ($\Delta\nu \approx 135\,\mu$Hz) and α Cen B ($\Delta\nu \approx 162\,\mu$Hz). The greyscale is the power spectrum of α Cen A as observed in a two-site campaign with UVES and UCLES (Bedding et al. 2004), but with weights optimized to minimize the sidelobes (Arentoft et al. 2010). The filled symbols are oscillation frequencies for the Sun (Broomhall et al. 2009; Table 2) after multiplying by 0.7816, and open symbols are frequencies for α Cen B (Kjeldsen et al. 2005) after multiplying by 0.6555. The scaling factors were tuned (using simple 'trial and error') to make the ridges for all three stars parallel. Note that scatter of observations about the smooth ridges for α Cen A and B is due to the relatively short duration of the observations (only a few times longer than the mode lifetimes).

There is a systematic progression in stellar parameters (mass, effective temperature and luminosity) as we go from α Cen A through the Sun to α Cen B. We see in Fig. 1 a corresponding progression in the positions of the ridges, which corresponds to a change in ϵ (see Equation 3). Apart from this, there is a close similarity between the three stars, although there are subtle differences in the curvatures of the ridges and in the small separations between them.

If two stars have very similar parameters, the ridges will almost coincide (Methods 1 and 2 become the same). This is the case for the Sun and the solar twin 18 Sco, recently observed with HARPS and SOPHIE (M. Bazot et al., in prep.), for which the scaling factor gives an extremely precise measurement of the mean density. In this case, we can also use one star as a guide when identifying modes in the other (and eliminating aliases). Another example is shown in Figure 2, for a pair of low-mass stars: the filled symbols show observed frequencies for τ Cet ($\Delta\nu \approx 170\,\mu$Hz; Teixeira et al. 2009) and the open filled symbols show those of α Cen B (Kjeldsen et al. 2005) after multiplying by 1.0478.

3.1. Measuring mean densities

Assuming that a given pair of stars are homologous, (i.e., that Equation 1 holds), the value of the scaling factor gives a direct measurement of their relative densities, without the need to refer to theoretical models. The scaling factors used in Fig. 1 are precise to about 0.05%, in the sense that changing them by this amount produces a noticable deviation from parallelism. If we accept the validity of Equation 1 then our results allow us to measure mean densities for α Cen A and B relative to solar with a precision of 0.1%. We obtain $(0.8601 \pm 0.0003)\,\mathrm{g\,cm^{-3}}$ for α Cen A and $(2.0018 \pm 0.0008)\,\mathrm{g\,cm^{-3}}$

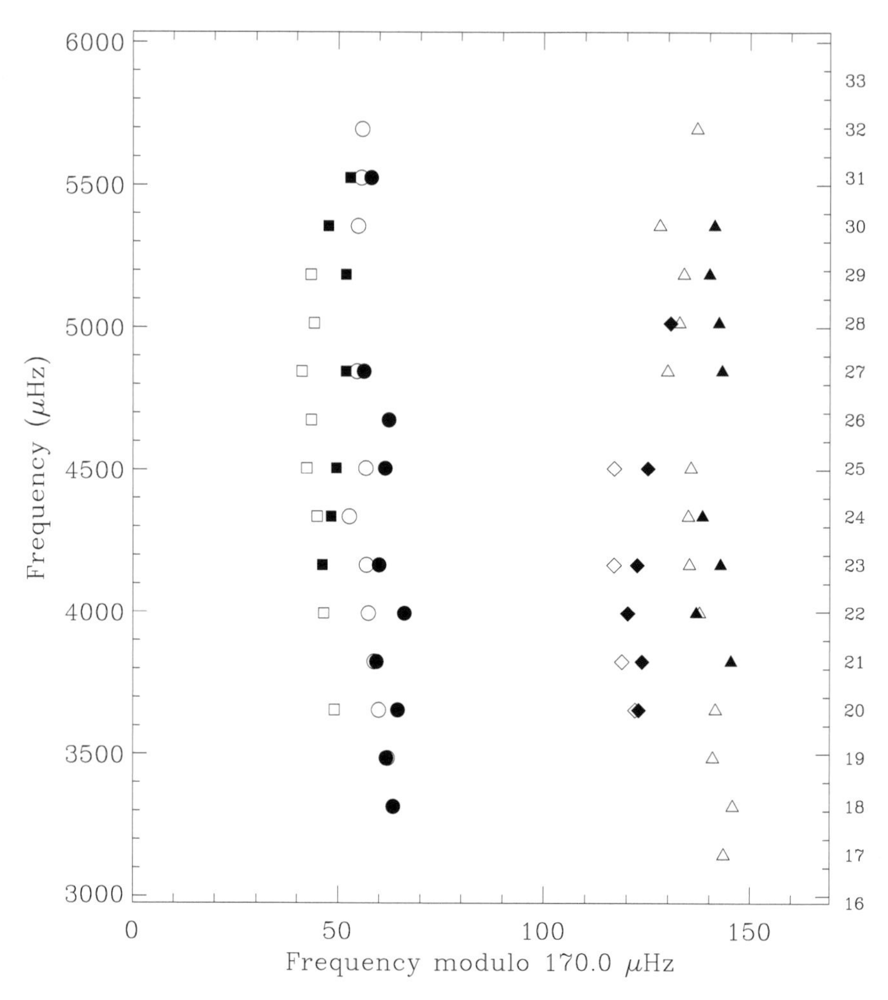

Figure 2: Scaled échelle diagram comparing two simliar low-mass stars. Filled symbols: frequencies (unscaled) for τ Cet. Open symbols: frequencies for α Cen B multiplied by 1.0478. The scaling factor was tuned to make the ridges coincide (Method 1/2). Symbol shapes indicate mode degree: $l = 0$ (circles), $l = 1$ (triangles), $l = 2$ (squares) and $l = 3$ (diamonds).

for α Cen B. These values agree with those found by comparing the observed frequencie of radial modes with models that have been corrected for the near-surface offset (Kjeldsen et al. 2008; Table 2), but are more precise and do not make any use of model calculations. In practice, however, Equation 1 may not be accurate to this level. In particular, defining a mean density requires that we specify the position of the surface, which is ambiguous, as discussed by Bahcall & Ulrich (1988) in the context of helioseismology. In any case, we have certainly derived mean densities for these stars that are precise enough for any practical application.

The two stars shown in Figure 2, τ Cet and α Cen B, are even closer. Once again, we found that changing the scaling factor by about 0.05% produced a noticable departure from parallelism. Using the mean density found above for α Cen B, the implied mean density for τ Cet is $2.198 \pm 0.004\,\mathrm{g\,cm^{-3}}$, which agrees with the value of $2.21 \pm 0.01\,\mathrm{g\,cm^{-3}}$ found by Teixeira et al. (2009) but is more precise. Again, we note that being able to measure the homology scaling factor to high precision does not necessarily provide a density measurement with similar accuracy.

3.2. Ridge identification in F stars

An important application of scaled échelle diagrams is the problem of ridge identification in F stars. This problem has arisen in the context of several F-type main-sequence stars observed using the CoRoT spacecraft. The first and best-studied example is HD 49933 ($\Delta\nu \approx 85\,\mu$Hz), whose échelle diagram from 60 days of CoRoT observations showed two broad and very similar ridges (Appourchaux et al. 2008). It was clear that one ridge was due to $l = 0$ and $l = 2$ modes and the other to $l = 1$, but the combination of significant rotational splitting and large linewidths made it difficult to decide which was which. Appourchaux et al. (2008) made a global fit to the line profiles, which led them to favor the possibility they labelled 'Scenario A'. Further analysis of the same data has been carried out by several groups (Benomar et al. 2009a; Gruberbauer et al. 2009; Roxburgh 2009) and none favored a definite identification, while comparison with theoretical models (Kallinger et al. 2010) gave a better match to Scenario B. Subsequently, the analysis of an additional 140 days of CoRoT observations using revised methods led Benomar et al. (2009b) to reverse the original identification in favor Scenario B.

Two other F stars observed by CoRoT have presented the same problem, namely HD 181906 ($\Delta\nu \approx 87.5\,\mu$Hz; García et al. 2009) and HD 181420 ($\Delta\nu \approx 75\,\mu$Hz; Barban et al. 2009). In neither case were the authors able to decide the correct scenario.

Using scaled échelle diagrams, together with the quite reasonable assumption that ϵ varies slowly with stellar parameters, we might hope to be able to tie these stars together. Figure 3 shows how this works for two CoRoT targets, HD 49933 and HD 181420. In all three panels, the filled symbols show Scenario B for HD 49933 (Benomar et al. 2009b). The open symbols show

Scenario 1 for HD 181420 (Barban et al. 2009) with three different scaling factors. In the upper panel, the scaling factor was chosen to align the ridges as closely as possible, again using trial and error, and we indeed see a good match. However, we should check whether shifting one star by half an order also produces a match. This requires changing the scaling factor by $0.5/n_{\mathrm{max}}$, which is about 2.5% in this case. This is shown in the lower two panels of Figure 3, where the scaling factor has been changed in both directions by this amount and then fine-tuned to align the ridges. Neither of these match as well as the upper panel, giving us confidence that Scenario B for HD 49933 is equivalent to Scenario 1 for HD 181420. That is, the two scenarios are either both correct or both wrong.

The third problematic CoRoT target mentioned above, which has a significantly lower signal-to-noise ratio, is HD 181906 (García et al. 2009). The power spectrum of this star is shown as the greyscale in Figure 4. Overlaid with filled symbols are the oscillation frequencies for HD 49933 from the revised identification (Benomar et al. 2009b; Scenario B) after multiplying by 1.011. There is good agreement between the stars and, using HD 49933 as a guide, we are able to follow the $l = 1$ ridge of HD 181906 down to quite low frequencies. Examining the two possible identifications proposed for HD 181906 by García et al. (2009), we can identify Scenario B for that star with Scenario B for HD 49933. Once again, either both are correct or both are wrong.

Having linked these three F-type CoRoT targets, all of which have quite similar values of $\Delta\nu$, we would clearly like to confirm the identifications by tying them to other stars whose identifications are secure. We do this in Figure 5. The greyscale shows the power spectrum of the CoRoT target HD 49385 ($\Delta\nu \approx 56\,\mu$Hz; Deheuvels et al. 2010), for which the $l = 0$ and 2 ridges are clearly resolved. The open symbols show frequencies for η Boo ($\Delta\nu \approx 40\,\mu$Hz) after multiplying by 1.390, with mode identifications that were verified by three sets of observations (Kjeldsen et al. 1995, 2003; Carrier et al. 2005) Finally, the filled symbols show once again the revised identification for HD 49933 (Benomar et al. 2009b; Scenario B), this time multiplied by 0.658. These scaling factors differ from unity by more than any others we have considered (since $\Delta\nu$ covers a bigger range). Despite this, we see good alignment of the ridges (Method 1) that gives a consistent picture. Interestingly, HD 49385 shows significantly more curvature at the lowest orders than the other two stars.

To summarise, we conclude that the correct identifications are: Scenario B for HD 49933 (Benomar et al. 2009b), Scenario 1 for HD 181420 (Barban et al. 2009) and Scenario B for HD 181906 (García et al. 2009). The first two of these agree with the conclusions of Mosser & Appourchaux (2009), which were based on autocorrelation analysis of the time series.

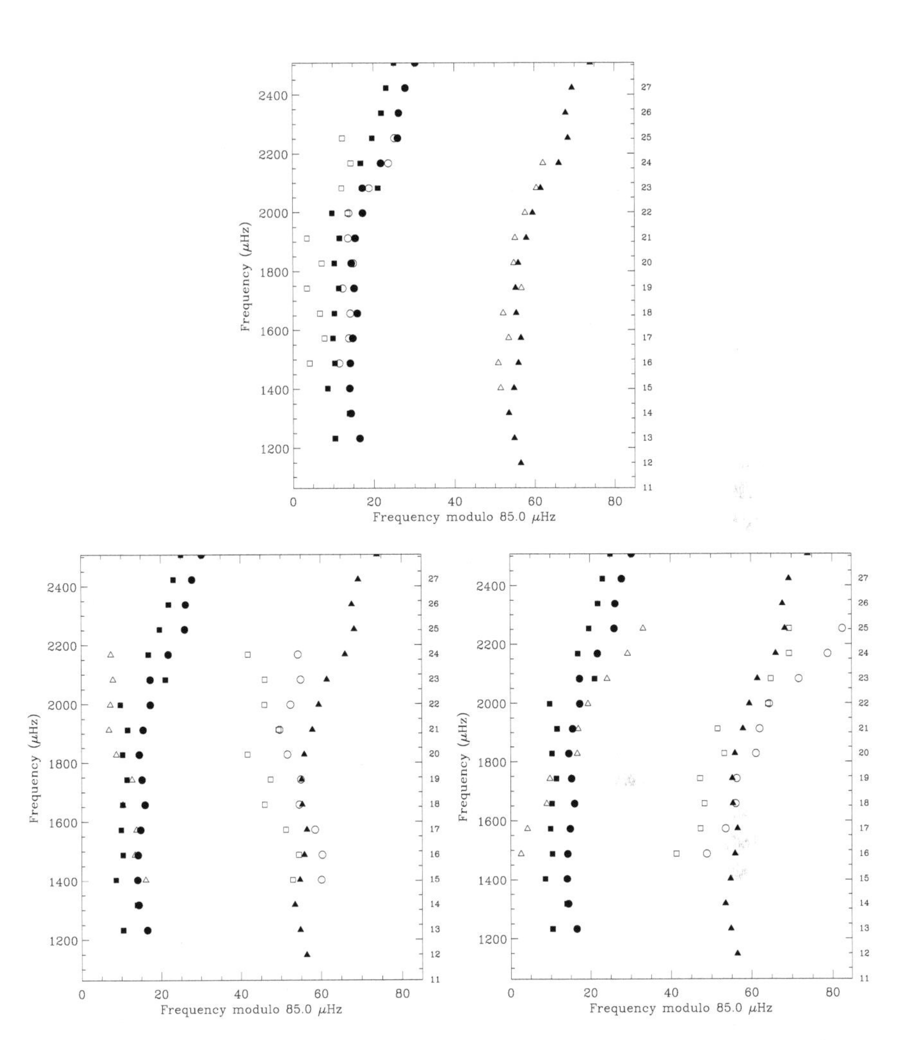

Figure 3: Scaled échelle diagrams showing two CoRoT F-type targets that have ambiguous identifications. The filled symbols are oscillation frequencies for HD 49933 from the revised identification (Benomar et al 2009b; Scenario B). The open symbols are frequencies for HD 181420 (Barban et al. 2009; Scenario 1) after multiplying by 1.144 (upper panel), 1.115 (lower left) and 1.173 (lower right). Symbol shapes indicate mode degree: $l = 0$ (circles), $l = 1$ (triangles), and $l = 2$ (squares).

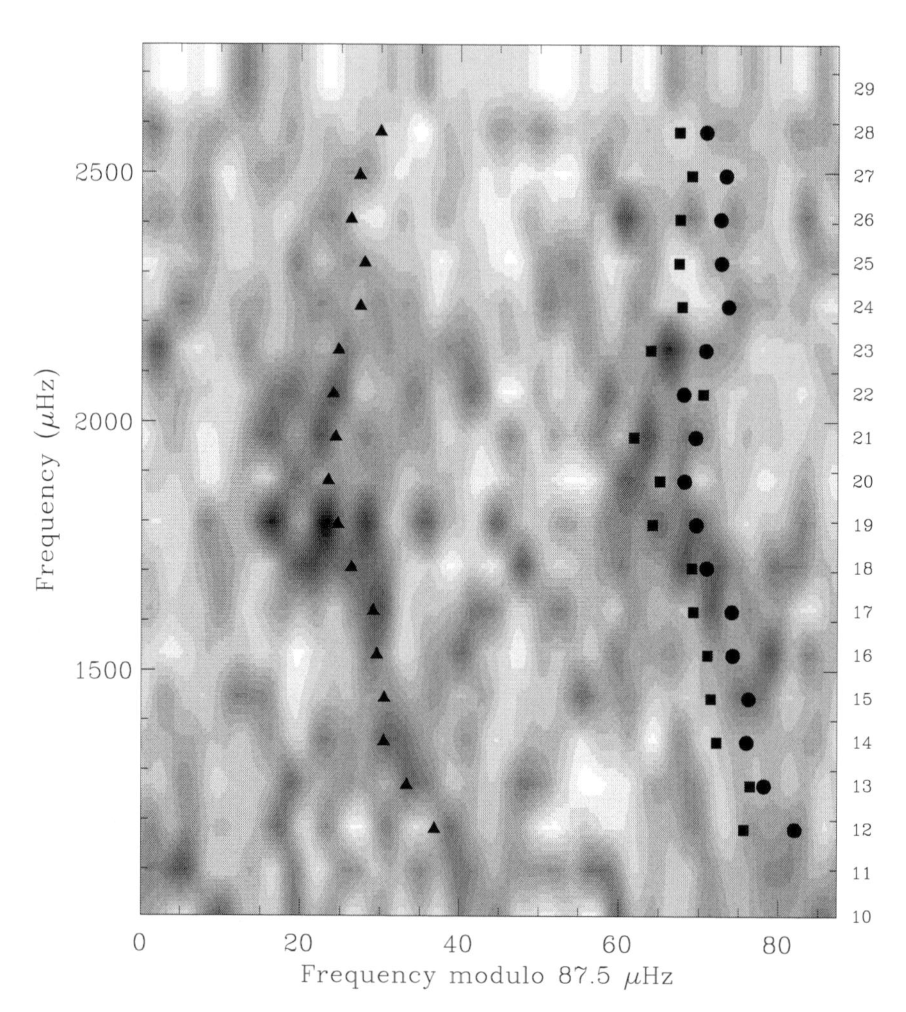

Figure 4: Scaled échelle diagram comparing HD 49933 with another CoRoT F-type target that has an ambiguous identification. The greyscale is the power spectrum of HD 181906 (García et al. 2009), smoothed to a FWHM of 3 μHz. The filled symbols are oscillation frequencies for HD 49933 from the revised identification (Benomar et al. 2009b; Scenario B) after multiplying by 1.011. Symbol shapes indicate mode degree: $l = 0$ (circles), $l = 1$ (triangles), and $l = 2$ (squares).

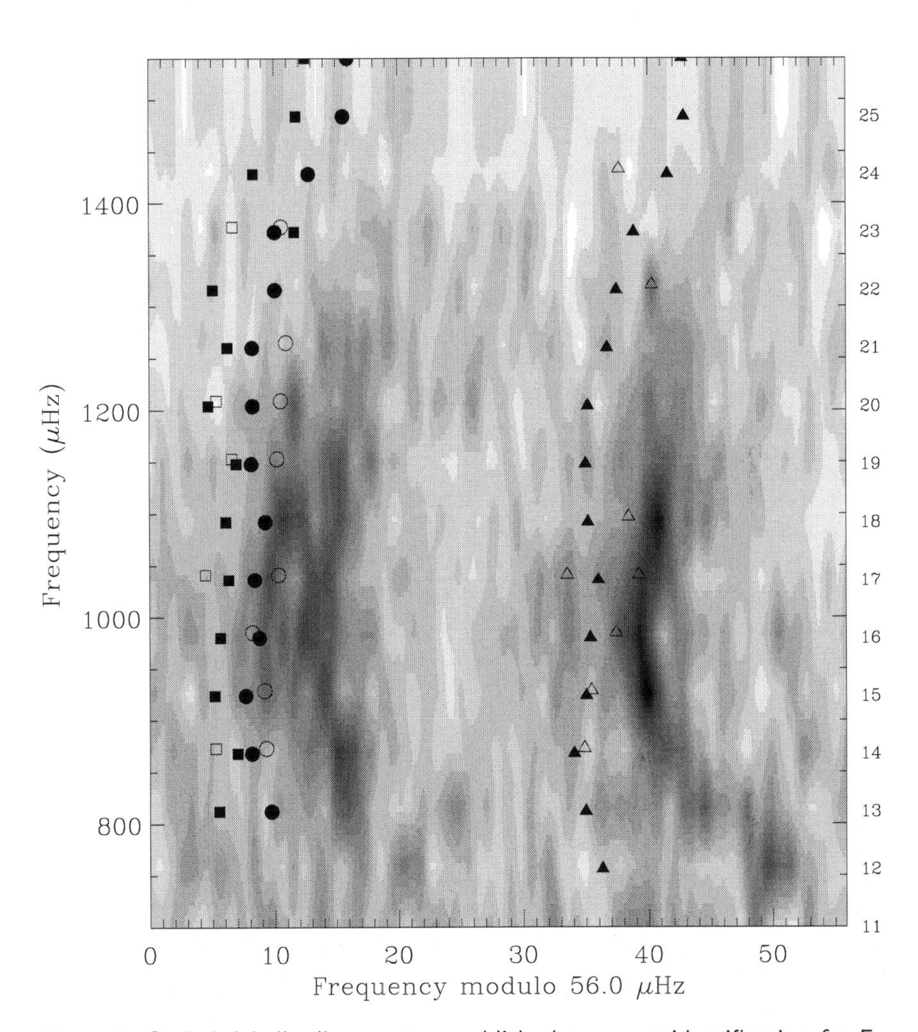

Figure 5: Scaled échelle diagram to establish the correct identification for F-type stars. The greyscale is the power spectrum of HD 49385 (Deheuvels et al. 2010), smoothed to a FWHM of $1\,\mu$Hz. The filled symbols are oscillation frequencies for HD 49933 from the revised identification (Benomar et al. 2009b; Scenario B) after multiplying by 0.658. Open symbols are frequencies for η Boo (Kjeldsen et al. 2003) after multiplying by 1.390. Symbol shapes indicate mode degree: $l = 0$ (circles), $l = 1$ (triangles), and $l = 2$ (squares).

4. Conclusions

We have described a method for scaling oscillation frequencies and displaying two or more stars on a single échelle diagram. Assuming that two stars are sufficiently similar to be homologous, the diagram can be used to infer the ratio of their mean densities very precisely, without reference to models. In addition, data from the star with the better signal-to-noise ratio can be used to confirm weaker modes and reject sidelobes in data from a second star. A very important application is to provide a solution to the problem of ridge identification in F-type stars observed by CoRoT, as discussed in Section 3.2., and we have successfully applied the method to Procyon (Bedding et al. 2010a). Another application is to apply what might be called ensemble asteroseismology to the very large samples of stars being observed by the CoRoT and Kepler space missions. The results of applying this technique to red giants observed with Kepler are described by Bedding et al. (2010b).

Acknowledgments. This work was supported financially by the Australian Research Council and the Danish Natural Science Research Council. We are very grateful to Rafael García for providing power spectra from CoRoT observations in electronic form, and to Sébastien Deheuvels, Eric Michel and colleagues for allowing us to show CoRoT results for HD 49385 in advance of publication. We thank Bill Chaplin and Dennis Stello for encouragement and useful discussions.

References

Appourchaux, T., Michel, E., Auvergne, M., et al. 2008, A&A, 488, 705

Arentoft, T., Kjeldsen, H., & Bedding, T. R. 2010, in GONG 2008/SOHO XXI Meeting on Solar-Stellar Dynamos asa Revealed by Helio- and Asterseismology, ed. M. Dikpati, I. Gonzalez-Hernandez, T. Arentoft, & F. Hill (ASP Conf. Ser.), in press (arXiv:0901.3632)

Bahcall, J. N. & Ulrich, R. K. 1988, Rev. Mod. Phys., 60, 297

Barban, C., Deheuvels, S., Baudin, F., et al. 2009, A&A, 506, 51

Bedding, T. R., Kjeldsen, H., Butler, R. P., et al. 2004, ApJ, 614, 380

Bedding, T. R. et al. 2010a, ApJ, submitted

Bedding, T. R. et al. 2010b, ApJ Lett., in press (arXiv:1001.0229)

Benomar, O., Appourchaux, T., & Baudin, F. 2009a, A&A, 506, 15

Benomar, O., Baudin, F., Campante, T. L., et al. 2009b, A&A, 507, L13

Broomhall, A.-M., Chaplin, W. J., Davies, G. R., et al. 2009, MNRAS, 396, L100

Brown, T. M. & Gilliland, R. L. 1994, ARA&A, 32, 37

Carrier, F., Eggenberger, P., & Bouchy, F. 2005, A&A, 434, 1085

Christensen-Dalsgaard, J. 2004, Sol. Phys., 220, 137

Christensen-Dalsgaard, J., Däppen, W., & Lebreton, Y. 1988, Nat, 336, 634

Deheuvels, S., Bruntt, H., Michel, E., et al. 2010, A&A, submitted

Dziembowski, W. A., Paternó, L., & Ventura, R. 1988, A&A, 200, 213

García, R. A., Regulo, C., Samadi, R., et al. 2009, A&A, 506, 41

Gough, D. O. 1986, in Hydrodynamic and Magnetodynamic Problems in the Sun and Stars, ed. Y. Osaki (Tokyo: Uni. of Tokyo Press), 117

Gough, D. O. 2003, Ap&SS, 284, 165

Grec, G., Fossat, E., & Pomerantz, M. A. 1983, Sol. Phys., 82, 55

Gruberbauer, M., Kallinger, T., Weiss, W. W., & Guenther, D. B. 2009, A&A, 506, 1043

Kallinger, T., Gruberbauer, M., Guenther, D. B., Fossati, L., & Weiss, W. W. 2010, A&A, in press (arXiv0811.4686)

Kjeldsen, H., Bedding, T. R., Baldry, I. K., et al. 2003, AJ, 126, 1483

Kjeldsen, H., Bedding, T. R., Butler, R. P., et al. 2005, ApJ, 635, 1281

Kjeldsen, H., Bedding, T. R., & Christensen-Dalsgaard, J. 2008, ApJ, 683, L175

Kjeldsen, H., Bedding, T. R., Viskum, M., & Frandsen, S. 1995, AJ, 109, 1313

Li, L. H., Robinson, F. J., Demarque, P., Sofia, S., & Guenther, D. B. 2002, ApJ, 567, 1192

Mosser, B. & Appourchaux, T. 2009, A&A, 508, 877

Rosenthal, C. S., Christensen-Dalsgaard, J., Nordlund, Å., Stein, R. F., & Trampedac, R. 1999, A&A, 351, 689

Roxburgh, I. W. 2009, A&A, 506, 435

Tassoul, M. 1980, ApJS, 43, 469

Teixeira, T., Kjeldsen, H., Bedding, T. R., et al. 2009, A&A, 494, 237

Comm. in Asteroseismology
Volume 161, May 2010

Photometric observations and frequency analysis of the δ Scuti star IP UMa*

D. Sinachopoulos[1], P. Gavras[1] and Chr. Ducourant[2]

[1]Institute of Astronomy and Astrophysics, National Observatory of Athens
I. Metaxa and Bas. Pavlou, GR-15236 Athens, Greece
[2]Observatoire de Bordeaux, 2 rue de l'Observatoire, BP 89
FR 33270 Floirac, France

Abstract

IP UMa is a δ Scuti star discovered by Hipparcos. 2642 observations of this target were acquired with the 20cm telescope of the Nea Lesbos Observatory in 10 nights from June 2 to July 9, 2009 using the Bessel V filter.

These data confirmed the pulsation frequency of the star listed in the Hipparcos Catalogue (10 c/d).

Accepted: May 25, 2010

Individual Objects: IP UMa

1. Introduction

IP UMa (HD 118954, BD +48 2141, HIP 66609) is listed in the revised catalogue of δ Sct stars by Rodriguez et al. (2000) as an A5 variable star with mean V=7.67, (B-V)= 0.32, amplitude 0.05 magnitudes and frequency 10 c/d. The relevant entry in this catalogue was based on information taken from Kazarovets et al. (1999), which introduces GCVS names for 3153 variable stars discovered by the Hipparcos mission. The data of the Rodriguez et al. catalogue were adopted from the corresponding Hipparcos catalogue entry (ESA 1997).

*Based on observations collected at the Nea Lesbos Observatory, Greece.

2. Observatory, Instruments, Observations

We searched for IP UMa pulsations by means of photometric observations, which were performed in ten photometric nights in the period June 2 - July 9, 2009 at the Nea Lesbos Observatory (38° 03' 14.1" N, 23° 49' 36.3" E, 270m altitude).

All data were collected using the 20cm VIXEN VISAC (f/9) telescope. Its SBIG 6303E science CCD camera has an array of 3060x2040 square pixels with dimensions 9x9 microns, corresponding to a surface of 27.5x18.4mm. The device is attached to the Cassegrain focus of the telescope, where each pixel corresponds to 1.03"x1.03" and the CCD array to a field of view (FOV) of 53'x35'. In this relatively large FOV one can usually find appropriate comparison and check stars. The temperature of the CCD was set to -10°C for lower DARK noise. Standard flat-fielding, BIAS and DARK corrections were always performed to the raw data.

All observations were autoguided using an ATIK 16IC CCD camera attached to a 5 inch Celestron telescope. This Schmidt-Cassegrain instrument has been mounted in parallel to the VIXEN 20cm. The two telescopes are attached to the same automatic SkyWatcher EQ6 SynScan Pro equatorial mount.

All devices are controlled via the (partially wireless) computer network of the observatory. Science CCD frames are automatically transferred to a dedicated workstation in the telescope warm-room. ESO-MIDAS software monitors the observing conditions and proceeds with the photometric data reduction.

We tried to position the stellar images to almost the same place of the CCD each night, in order to minimize systematic effects of the instrumental photometry. Observations were terminated when an airmass of 2.0 was reached. Exposure time was usually 18 seconds, but 10 seconds was also used. 2642 measurements of the field centered on IP UMa were collected, using the Bessel V filter.

The median of the seeing conditions is 2.4 arcseconds, corresponding to a well sampled stellar PSF, which is needed for accurate photometry. The minimum observed seeing was 1.2". In such cases we were defocusing the telescope slightly, so that the stellar images come to a FWHM of two pixels or more. Table 1 gives a journal of observations providing an overview of HJD, the total number of data points collected and photometric hours per night.

3. Comparison and Check Stars

There were only few suitable candidates for comparison and check stars in the FOV. We chose HIP 66713 (V=8.37, (B-V)=0.39) as comparison and HIP 66485 (V=8.57, (B-V)=1.01) as check star for obvious reasons.

Table 1: Journal of Observations

HJD [+2400000]	DURATION [Hours]	Exposures No
54985	3.3	194
54987	6.0	385
54991	4.5	295
54993	5.2	314
54995	5.5	327
54997	5.1	372
54998	2.6	87
55020	2.4	185
55021	2.3	231
55022	3.1	252

We checked these two stars for variability performing their relative photometry each night. The stars were not variable within the observational run of 38 days. We performed a linear regression to the results of each night. In this way we checked whether the magnitude difference between comparison and check stars changes systematically with airmass due to extinction effects caused by their colour difference. The slopes of the linear regressions were always found to be statistically zero with a mean R.M.S. error of 0.013 magnitudes. The mean magnitude difference of the two stars at the zenith was $\Delta V = 0.243 \pm 0.002$ magnitudes.

4. Frequency analysis

Once the observations were reduced and magnitude differences of each exposure determined, we applied a mean filter to the results with a size of four observations, before proceeding to the frequency analysis.
At a first glance the light curves show moderate variations from night to night. The largest peak-to-peak amplitude is 0.13 mag. The Fourier analysis was performed using Period04 (Lenz & Breger 2005). The frequency resolution of this run is 0.04 c/d (Loumos & Deeming 1978), or lower. The spectral window is shown in Figure 1a.

After the Fourier analysis of the magnitude differences ΔV, the frequencies, amplitudes and phases were improved by a least squares fit. The first frequency found (f_1) is the one detected by Hipparcos (10.0 c/d) as shown in Figure 1b. A search for a second frequency in the prewhitened data was performed, revealing

10.95 c/d. The $(1\ \text{day})^{-1}$ ambiguity caused by the 1 c/d aliasing is shown in Figure 1c. After fitting f_1 and f_{1a} the sigma of the residuals has a standard deviation of 0.0096 magnitudes.

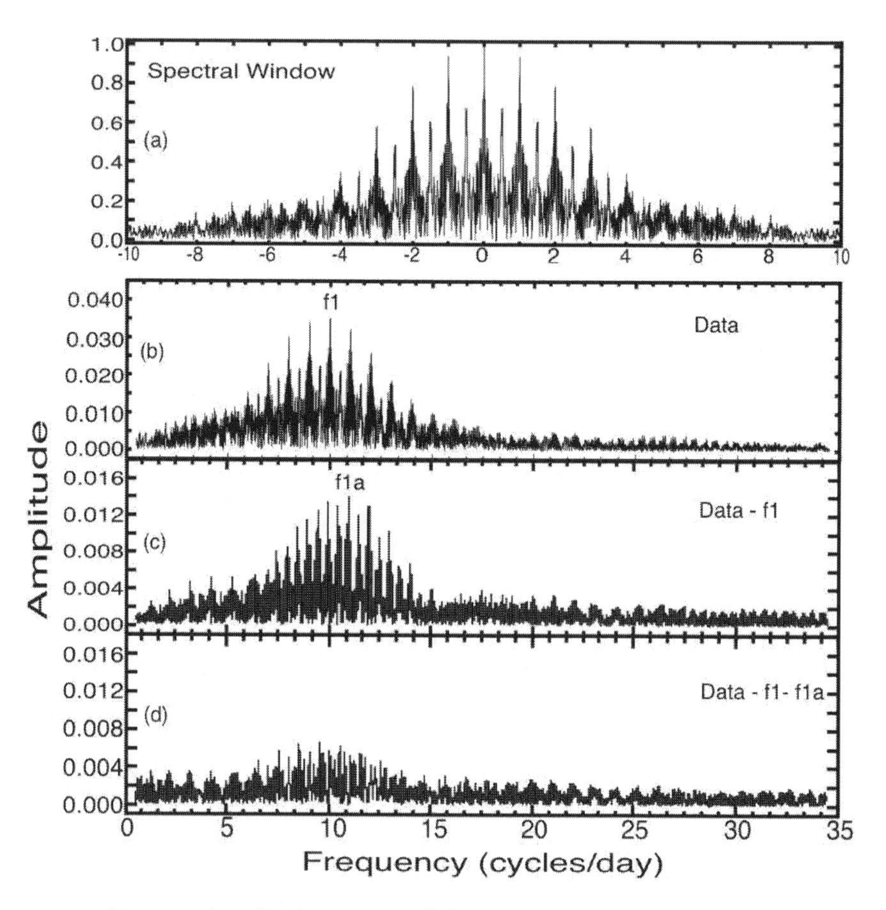

Figure 1: Amplitude spectra of the June 2 - July 9, 2009 data

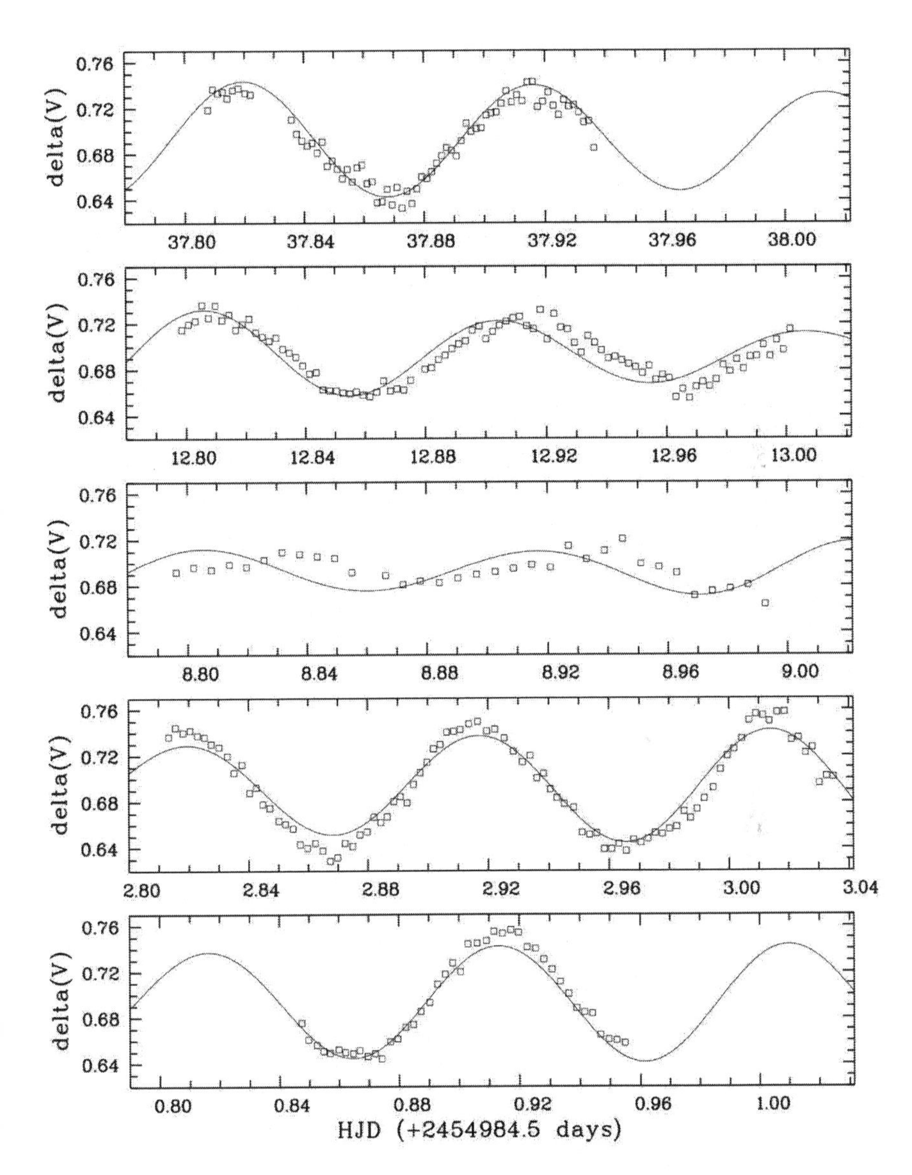

Figure 2: Observations of five selected nights of the observations of IP UMa carried out in 2009, together with the fit of the frequencies found

Table 2: Computed frequency for IP UMa (V-data) and its alias

Id.	Frequency [c/d]	Semi-ampl. [mag]	Phase [cycles]	S/N
f_1	9.9972	0.0340	0.206	14
σ_{f_1}	0.0003	0.0007	0.003	

Table 2 contains the characteristics of frequency f_1. The first line contains the frequency, its semi-amplitude, phase and signal to noise ratio. The second one lists uncorrelated uncertainties of the corresponding values in line one as calculated by the least-squares-fit.

Figure 2 shows data of five selected nights of our observations of IP UMa. Heliocentric Julian date is shown in the abscissa with an offset of -2,454,984.5 days, which corresponds to midnight (UT) June 02, 2009 CE. The magnitude difference, plotted in the ordinate, is the one between comparison (HIP 66713) and IP UMa. The f_1 and f_{1a} are also shown fitted to the data.

Acknowledgments. The authors are very thankful for the comments of the anonymous referee.

References

ESA, 1997, The Hipparcos and Tycho Catalogues, ESA SP-1200

Kazarovets, A. V., Samus, N. N., Durlevich, O. V., et al.1999, Information Bulletin on Variable Stars, 4659, 1

Lenz, P., & Breger, M. 2005, CoAst, 146, 53-136

Loumos, G.L., & Deeming, T.J. 1978, ApSS, 56, 285

Rodriguez, E., Lopez-Gonzalez, M. J., & Lopez de Coca, P. 2000, A&AS, 144, 469-474

Comm. in Asteroseismology
Volume 161, June 2010

Photometric and Spectroscopic Study of the δ Scuti Stars FH Cam, CU CVn and CC Lyn

G. J. Conidis[1,2], K. D. Gazeas[3],
C. C. Capobianco[4], W. Ogloza[5]

[1]Department of Physics and Astronomy, York University, 128 Petrie Science and Engineering Building, 4700 Keele St., Toronto, Ontario, M3J 1P3, Canada
[2] David Dunlap Observatory, University of Toronto, P.O. Box 360, Richmond Hill, Ontario, L4C 4Y6, Canada
[3] Department of Astrophysics, Astronomy and Mechanics, Faculty of Physics, University of Athens, GR-15784 Zografos, Athens, Greece
[4] Department of Physics, Engineering Physics and Astronomy at Queen's University, Kingston, ON, K7L 3N6, Canada
[5] Mt. Suhora Observatory of the Pedagogical University, ul. Podchorazych 2, 30-084 Cracow, Poland

Abstract

Three short period (P $\ll$ 1 day) variable stars from the Hipparcos catalogue targets were observed after suspected misclassification as β Lyr eclipsing systems (Perryman et al. 1997), as no secondary component had been noticed in the inspection of their Broadening Functions (BFs) (Rucinski 2002). FH Cam is found to be a multiple star system with a member exhibiting δ Scuti behaviour. The dominant pulsation frequency is found to be 7.3411 $\pm$ 0.0002 c/d, which corresponds to a pulsation mode of $\ell \leq 1$. We confirmed the pulsations of CU CVn using photometric observations and found a pulsation frequency of 14.7626 $\pm$ 0.0250 c/d, which is in agreement with the period given in literature. CC Lyn is a non-eclipsing visual binary (CCDM J07359+4302AB), the brighter component (A) is found to be a multi-mode δ Scuti pulsator, with pulsation frequencies of 5.6402 $\pm$ 0.0004 c/d and 7.3368 $\pm$ 0.0005 c/d.

Accepted: June 7, 2010

Individual Objects: FH Cam, CU CVn, CC Lyn

1. Introduction and Motivation

Among the contact binary candidates, within the David Dunlap Observatory (DDO) Contact Binary Survey (Rucinski et al. 1999; Pribulla et al. 2009), three candidates FH Cam, CU CVn and CC Lyn (CCDM J07359+4302A) had no sign of a spectroscopic companion. Since these Hipparcos variable stars, have variability periods significantly shorter than half a day, it was conjectured they might be δ Scuti pulsators (Rucinski 2002).

The purpose of this paper is to confirm the pulsational nature of these variable stars, calculate their pulsational frequencies and, when possible, identify their pulsational harmonic modes. This was performed by using both photometric and spectroscopic observations, which are discussed right after.

2. Observations and Data Reductions

For the spectroscopic observations, we used the 1.88 m Cassegrain telescope at the DDO, which was outfitted with a Thomson Photometrics CCD detector (1024$\times$1024 pixels, pixel size = 19.0$\times$19.0 μm) during the 2000 – 2001 observing runs. The Cassegrain spectrograph scale at the dispersion of 10.8 Åmm^{-1} is about 0.2 Åpixel^{-1} or about 12 km s^{-1} pixel^{-1}. All the spectra collected using the 2000 – 2001 configuration were taken at a center wavelength of 5185 Å (which corresponds to approximately having the Mg I triplet centered), giving a spectrum coverage of 210 Å, on the 1800 lines mm^{-1} grating. The 2004 – 2005 observations had been taken with a Jobin-Yvon CCD Camera (2048$\times$512 pixels, pixel size = 13.5$\times$13.5 μm), with a center wavelength of 6440 Å, and the same grating. The change in wavelength was due to a suspicion of flexure problems with the spectrograph, so it was decided to select a spectral window which included telluric features from which the flexure of the telescope could be monitored and the residual shift in radial velocity space could be removed for each spectra. It should be noted that spectra collected during March 2005 (five nights) and April 16, 2005 (one night) were badly corrupted by flexure problems from the spectrograph which imposed a random additive constant to the calculated radial velocities after every comparison spectra. All these spectra were discarded and not used in the radial velocity analysis of this paper. However, these aforementioned dates and the number of collected spectra and images are still listed in Table 1, since the photometric data collected in parallel were used for the analysis. During all observing runs, wavelength calibration spectra of an Fe-Ar lamp were taken frequently, depending on the length of the individual exposures. Spectra of multiple template (radial velocity standard) stars, with similar spectral type, were collected at the beginning and the end of each night. The 2004 – 2005 observing runs vary in that telluric standards were also acquired during the observing run, at the beginning or the end of the night. The spectroscopic data were bias and dark corrected, rectified and wavelength calibrated using the IRAF software package. The procedure of cosmic ray removal was done using a separate, standalone program (Pych 2003). The spectra were then extracted into one-dimensional spectra using IRAF. The radial velocities were calculated using the broadening function (BF) algorithms (Rucinski 1992). The BF method was applied to the entire spectral range of the collected spectra, which has been shown to be a viable method to detect non-radial pulsations (Pribulla et al. 2009). Line profile variation detection was not a possible method of detecting frequencies since the S/N of the spectra were $\sim$ 12-15. One should bear in mind that the CCD on the 1.88 m had been altered multiple times throughout the time spanned by the observations (2000 - 2005), giving an inhomogeneity to the quality of data which induces a slight change in the error bars. This shows up as slightly scattered data taken with center wavelength 6440 Å, relative to the data centered at 5185 Å.

Date	Star	Images	Spectra (S/N)	Template	Telluric Std.	Obs.
21/01/2000	CU CVn	0	19 (15)	HD 102870	N/A	DDO
24/04/2000	FH Cam	0	46 (15)	HD 48843	N/A	DDO
24/04/2000	CU CVn	0	12 (15)	HD 102870	N/A	DDO
21/02/2001	CC Lyn	0	27 (15)	HD 32963	N/A	DDO
22/02/2001	CC Lyn	0	36 (15)	HD 32963	N/A	DDO
19/12/2004	CC Lyn	0495	74 (13)	HD 103095	HD 87901	DDO
28/12/2004	CC Lyn	0273	64 (13)	HD 103095	HD 87901	DDO
06/03/2005	CC Lyn	0024	0 (0)	N/A	N/A	DDO
12/03/2005	CC Lyn	0078	22 (12)	HD 114579	HD 87901	DDO
26/03/2005	CC Lyn	0071	16 (12)	HD 196821	HD 87901	DDO
26/03/2005	CU CVn	0187	13 (12)	HD 196821	HD 159139	DDO
26/03/2005	FH Cam	0026	0 (0)	N/A	N/A	DDO
09/04/2005	FH Cam	0177	49 (12)	HD 128167	HD 136849	DDO
16/04/2005	FH Cam	0070	29 (12)	HD 128167	HD 136849	DDO
14/04/2006	FH Cam	0037	0	N/A	N/A	UAO
03/05/2006	FH Cam	2948	0	N/A	N/A	UAO
06/05/2006	FH Cam	4758	0	N/A	N/A	UAO
07/05/2006	FH Cam	4997	0	N/A	N/A	UAO
22/05/2006	FH Cam	1241	0	N/A	N/A	UAO

Table 1: Log of the type and amount of data acquired on a specified evening for a given target at the indicated observatory

The photometric data gathered at the DDO are collected using a 0.15 m f/8 refractor, attached in parallel on the 1.88 m telescope for guiding purposes. This refractor is outfitted with a SBIG ST-6 CCD Camera (375×241 pixels, pizel size

= 8.6×6.5 μm), with a field of view of 15×10 arcmin. Dark, bias, flat frames were acquired in every observing night. The reduction of the photometric exposures were done with in AIP4WIN (Berry & Burnell 2000). The CCD exposures were taken without a filter. (Table 2 lists the comparison and check stars for each target-object). Acquiring a check star for CU CVn was not possible as the field of view with this setup was too small to allow three bright stars to be simultaneously monitored.

Further observations of FH Cam were carried out from the University of Athens Observatory (UAO), in Athens, Greece with a 0.40 m f/8 Cassegrain telescope. The observations were collected over 5 nights, in April and May 2006. For the observations obtained on April 14 and May 22, 2006 a SBIG ST-8 CCD detector (1530×1020 pixels, pixel size = 9.0×9.0 μm) was used, giving a field of view of 15×10 arcmin. For the observations obtained on May 3, 6 and 7, 2006 an ST10-XME CCD detector (2184×1472 pixels, pixel size = 6.8×6.8 μm) was used, giving a field of view of 12×11 arcmin. The star was observed in most of the nights for at least 5 hours (typically 5-7 hours), covering more than one period in Bessel B, V, R and I filters. The CCD frames were bias, dark and flat field corrected, using bias, dark and flat images obtained at the beginning or end of each night. A complete list of photometric and spectroscopic observational logs are summarized in Table 1.

3. Overview of Targets

Table 2 lists the coordinates, photometric magnitude, spectral classification, the projected stellar-surface equatorial rotation velocity, as well as the comparison and check stars used for the photometric reduction. In addition, it lists the previously published pulsating frequencies and spatial radial velocities for all three targets. FH Cam was classified as a β Lyr eclipsing binary system by the light curve obtained by the Hipparcos mission (Perryman et al. 1997). Similarly, CU CVn was initially classified as a β Lyr eclipsing binary system by the Hipparcos mission (Perryman et al. 1997), but in 2002 it was discovered to be a δ Scuti pulsator (Vidal-Sainz J. et al. 2002). CC Lyn (HD 60335) is the brighter component of the visual binary system CCDM J07359+4302AB ($m_A = 6.538$, $m_B = 8.257$, $PA = 88^o$, $Sep = 2.2''$).

4. Data Analysis

After converting the dates for the observations to Heliocentric Julian Date (HJD), the photometric data were examined for erroneous outlying points. If data points were found to be outside the 3σ amplitude range and showed no temporal-continuity with adjacent points indicated a possible physical origin,

	FH Cam	CU CVn	CC Lyn
R.A.	07 57 39.7509	13 48 20.1169	07 35 55.9777
Dec.	+77 34 35.121	+31 24 03.801	+43 01 51.453
m_v	6.903	7.537	6.538
Sp. Class	A8V	A7V	F0-2V
$v \sin i$ (km/s)	45	155	≤ 15
Comparison star	GSC 4531-1441 (HD 63311)	GSC 2537-0397 (HD 120279)	GSC 2966-0661
Check star	GSC 4543-1724 (HD 63855)	N/A	GSC 2966-0781
ν_{pulse} (c/d)	3.67002(1)(HIP)	7.3709(8)(HIP) 14.7418(6) † 12.5612(3)†	2.81990(4)(HIP)
$RV_\circ$ (km/s)	3.0 4.5 ± 7.1††	-4.1	Variable 20.6 ± 2.0‡

Table 2: Overview of basic information of the three observed targets. The Comparison and Check stars were used for the photometric reduction. Spectral types, *v*sin*i* and radial velocities were taken from Pribulla et al. 2009 or otherwise mentioned. Previously published pulsating (or orbital) frequencies and spatial radial velocities of targets are taken from Hipparcos catalog (pulsating frequencies are double the orbital frequency mentioned by HIP) and † Vidal-Sainz J. et al 2002, †† Grenier et al. 1999, ‡ Wilson 1953

the points were discarded. The remaining time series was then analyzed using the software package Period04 (Lenz & Breger 2005). This package was used to calculate the Discrete Fourier Transforms of the data, enabling a determination of the oscillation frequencies. The errors were calculated using the Monte Carlo error estimation package within Period04, where 1000 iterations were used for all the photometric data. The signal to noise ratio (S/N) was also calculated by Period04. The resulting frequencies and their signal to noise ratio are both listed in Table 3. The amplitude spectra of all stars are presented for each frequency in Figure 5. In order to attain the phase plots for the photometric data, the location of the photometric light curve minima of the pulsations T_{min} was identified. In the case of FH Cam, the B filter photometry was used to identify T_{min} and was used to create all phase plots in the V, R and I filters. In the cases of CU CVn and CC Lyn, the unfiltered photometry was used to determine T_{min}.

In the case of the photometric data taken from the UAO, the data was of very good quality ($\sigma_{mag} \approx 1$ mmag) and was acquired in the B, V, R, I Bessel passband filters. This enabled the use of the software package FAMIAS (Zima 2008), which requires multi-passband filtered data to identify the spherical harmonics associated with detected frequencies.

Photometric				
Star	ν_{pulse} (c/d)	Amplitude (mag)	N	S/N
FH Cam[†]	07.3411 ± 0.0002	0.0313 ± 0.0003	3808	4.60
CU CVn	14.7626 ± 0.0250	0.0300 ± 0.0010	0284	4.81
CC Lyn ν_1^*	05.6402 ± 0.0004	0.0245 ± 0.0010	0776	5.49
CC Lyn ν_2^*	07.3368 ± 0.0005	0.0123 ± 0.0010	0776	3.38
Spectroscopic				
Date	$RV_\circ$ (km/s)	Amplitude (km/s)	N	
FH Cam				
24/04/2000	+03.21 ± 0.17	2.78 ± 0.14	0050	
09/04/2005	-06.88 ± 0.10	1.64 ± 0.24	0049	
CU CVn				
All Spectra	-00.93 ± 0.17	1.49 ± 0.42	0032	
CC Lyn				
21/02/2001	+17.09 ± 0.16	2.52 ± 0.20	0027	
22/02/2001	+12.83 ± 0.12	1.94 ± 0.19	0036	
19/12/2004	+33.87 ± 0.28	3.11 ± 0.38	0074	
28/12/2004	+51.30 ± 0.23	6.25 ± 0.32	0064	

Table 3: A listing of the pulsating frequencies and their corresponding amplitudes, found in our analysis, as well as the associated velocity amplitudes and phases found from folding the RV data to the same periods. N is the number of exposures used to calculate the period listed. [†] The photometric data used for FH Cam was acquired through a B filter, while the remaining two stars had their photometric data collected through clear apertures.

Similarly, the radial velocity data was converted to HJD and phase plots where generated using the photometrically determined frequencies for each star. The value T_{min} used for the radial velocity phase plot was the same used for the corresponding stars photometric plot. The phased radial velocity data was then fitted with a sinusoidal curve, to determine the amplitude of the radial velocity variations, and the radial velocity offset ($RV_{\circ}$) (Table 3).

4.1. FH Cam

From the analysis of the B filter photometric data, the dominant period was determined to be 7.3411 $\pm$ 0.0002 c/d (see Figure 5), which is in agreement with double the Hipparcos frequency (7.34004 $\pm$ 0.00002 c/d). The amplitude of the photometric oscillation decreases as the filter's transparent window moves toward longer wavelengths, which supports the classification of FH Cam being a pulsator and not an eclipsing binary (See Figure 1 and Table 4).

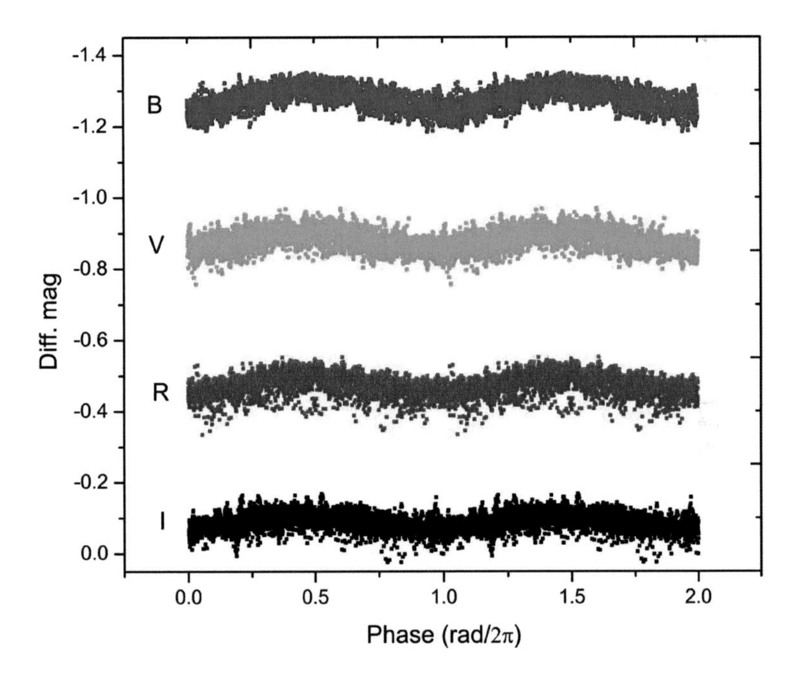

Figure 1: Bessel passband photometry of FH Cam. The decrease in the amplitude of the oscillation, in the order of B, V, R, I passband filters (Top to Bottom) clearly indicates a pulsation is the cause of the change in magnitude. The pulsation shown has a frequency corresponding to 7.3411 c/d.

Filter	Amplitude (mmag)	Phase (rad $(2\pi)^{-1}$)
B	30.97 ± 0.37	0.762 ± 0.002
V	23.73 ± 0.37	0.717 ± 0.003
R	22.87 ± 0.42	0.715 ± 0.003
I	17.63 ± 0.44	0.697 ± 0.005

Table 4: The amplitude and phase from fitting a sinusoidal function to the data in Figure 1.

The software package FAMIAS was used to calculate models simulating the frequencies found from the Bessel B, V, R, I photometry in order to constrain the radial order ℓ. The assumptions regarding the models of FH Cam are the following: $M = 2.0M_{\odot}$, $T_{eff} = 7650 \pm 100$ K and $\log(g) = 4.0 \pm 0.1$, which are the average values of 10 stars with a spectral classification of A8V. A grid within the Kurucz atmospheric models were found to fit the data, with a range of metallicities of $0 < Z < 0.02$ and micro turbulence ranging 0 km/s $< \xi <$ 4 km/s with no noticeable change to the values of ℓ outputted by the software package. The modal orders $\ell = 0$ to 5 were found to agree for all the amplitude ratio vs. phase difference plots except the R and V amplitude ratio vs. phase diagram (see Figure 6) which shows that only the $\ell = 0$, 1 radial pulsations are allowed. Hence, we find that the detected pulsation must have an order of $\ell \leq 1$.

The systemic velocity ($RV_{\circ}$) and the amplitude of the pulsation, are variable on a nightly basis (See Table 3 and Figure 2), which indicates that FH Cam is a multiple system with a non-spectroscopic and non-visual companion. While the time-dependent amplitudes of the radial velocity pulsations are most likely the result of other excited (low amplitude) pulsation frequencies constructively or destructively interfering with the observed pulsation.

4.2. CU CVn

From the analysis of the unfiltered photometric data, the dominant frequency was determined to be 14.7626 ± 0.0250 c/d (see Figure 5), which is in complete agreement with the spectroscopic data (See Figure 3), the dominant frequency found by Vidal-Sainz et al. 2002 (14.7418 ± 0.00004 c/d), and the double-frequency from Hipparcos (14.7418 ± 0.00001 c/d[1]).

[1]Hipparcos assumed CU CVn was a binary, so multiplication by two is needed to obtain the correct frequency.

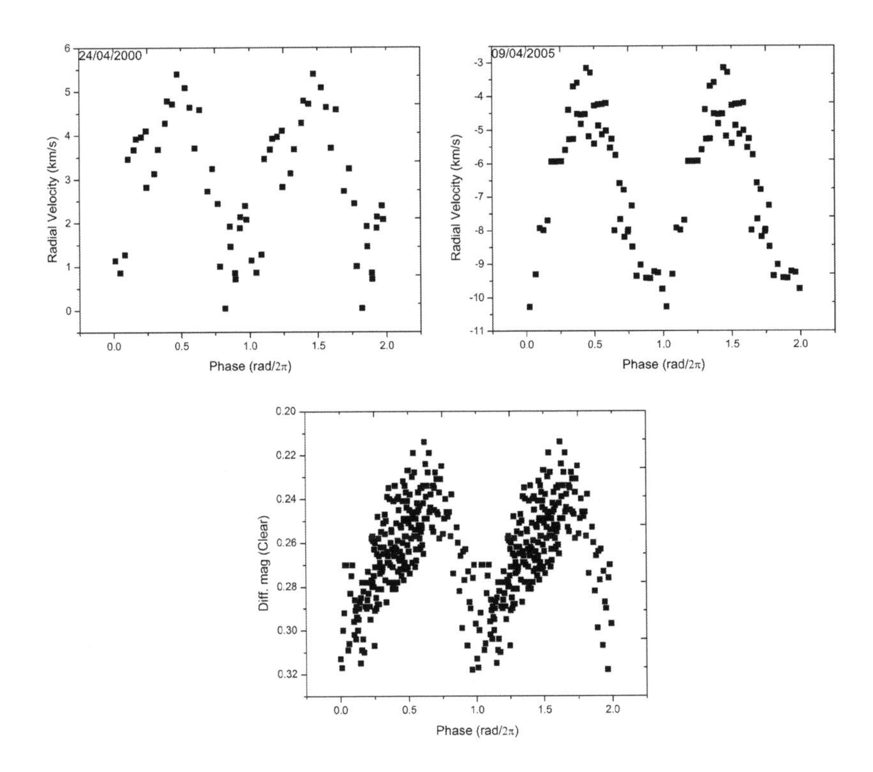

Figure 2: Radial velocity and differential magnitude phase plots for FH Cam, for the data acquired at the DDO. The frequency of pulsation is 7.3411 c/d.

The systemic velocity was found to be -0.93 ± 0.17 km/s for all the spectra acquired during the months of January to March in the year 2000.

4.3. CC Lyn

From the analysis of the unfiltered photometric data (See Figure 4), the dominant frequency was determined to be 5.6402 ± 0.0004 c/d (see Figure 5), which is also in agreement with the double-frequency from Hipparcos (5.63980 ± 0.00002 days). This is also in strong agreement with the radial velocity data being folded about this frequency (Figure 4). Interestingly, the ratio of the dominant (largest amplitude) frequency (ν_1^*) to the second dominant frequency (ν_2^*) gives a value of 0.768 ± 0.005, which is in agreement with the value 0.77 for the ratio of first radial overtone pulsation to the fundamental

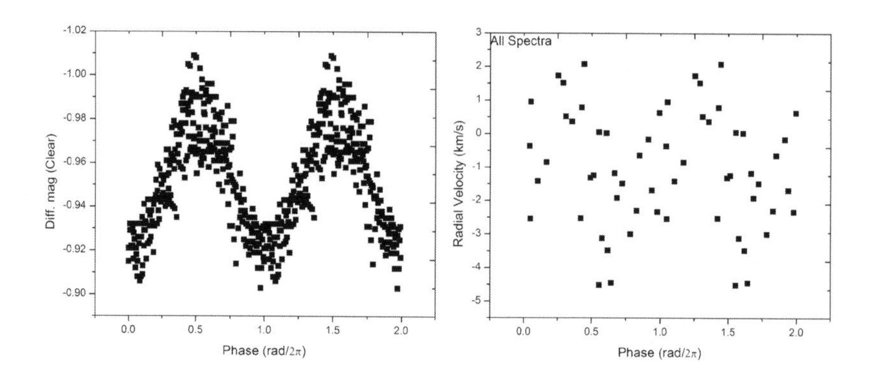

Figure 3: Radial velocity and differential magnitude phase plots for CU CVn for data acquired at the DDO, for the found frequency of 14.7626 c/d.

radial mode pulsation (Kurtz 2006). The value of this ratio could be an indication that CC Lyn is a Pop I star (Breger 2000), which has a frequency ratio of 0.769 ± 0.001. The aforementioned agreement with the expected ratio for a double-mode δ scuti pulsator must be confirmed through mode id calculations to properly secure its classification as a double-mode pulsator.

CC Lyn also shows variations in its systemic radial velocity and radial velocity pulsation amplitude and phase, as indicated in Table 3. CC Lyn is known to be in a binary system, which we confirmed through the observed change in the systemic velocity in each observing run. The variational change in amplitude and phase is expected in a multi-mode pulsator, as the different pulsations will beat against each other creating constructive and destructive interference, observed as a modulating pulsation.

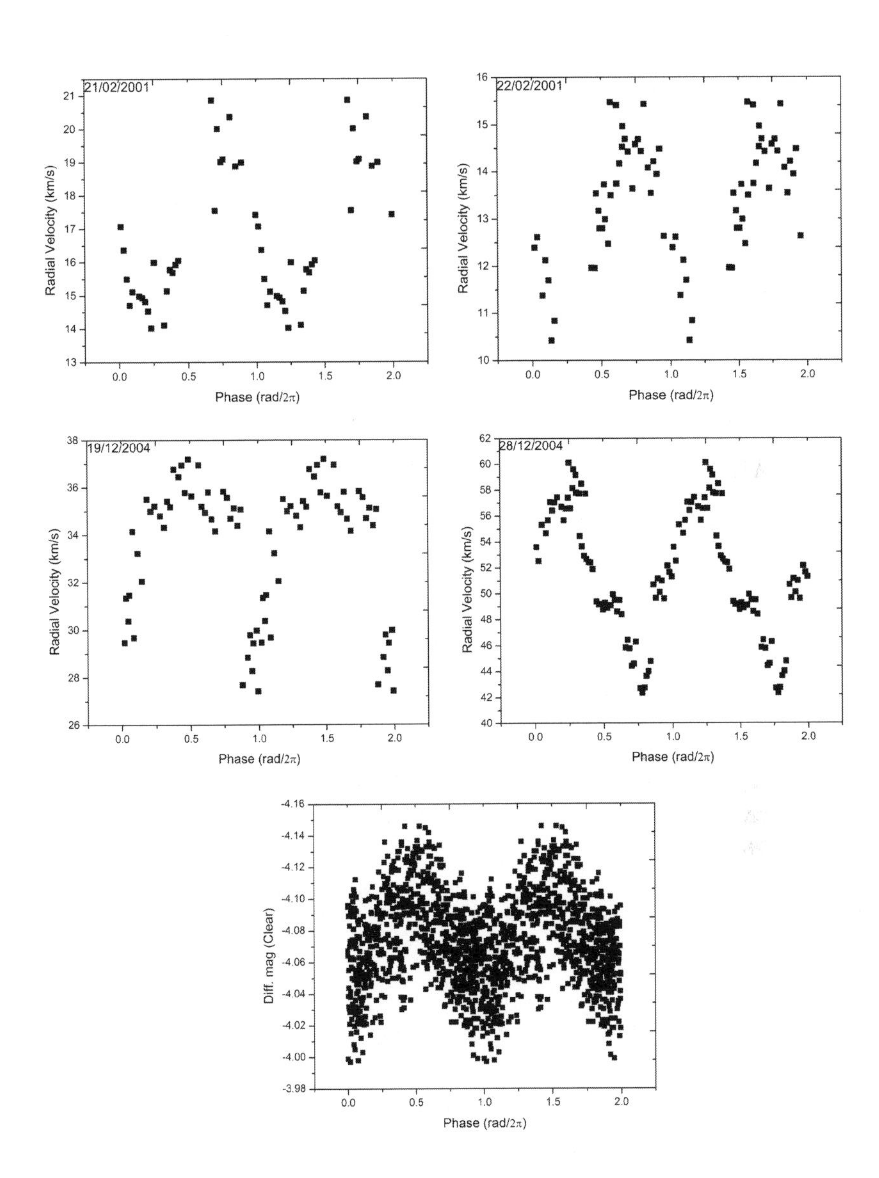

Figure 4: Radial velocity and differential magnitude phase plots for CC Lyn, for the frequency of 5.6402 c/d.

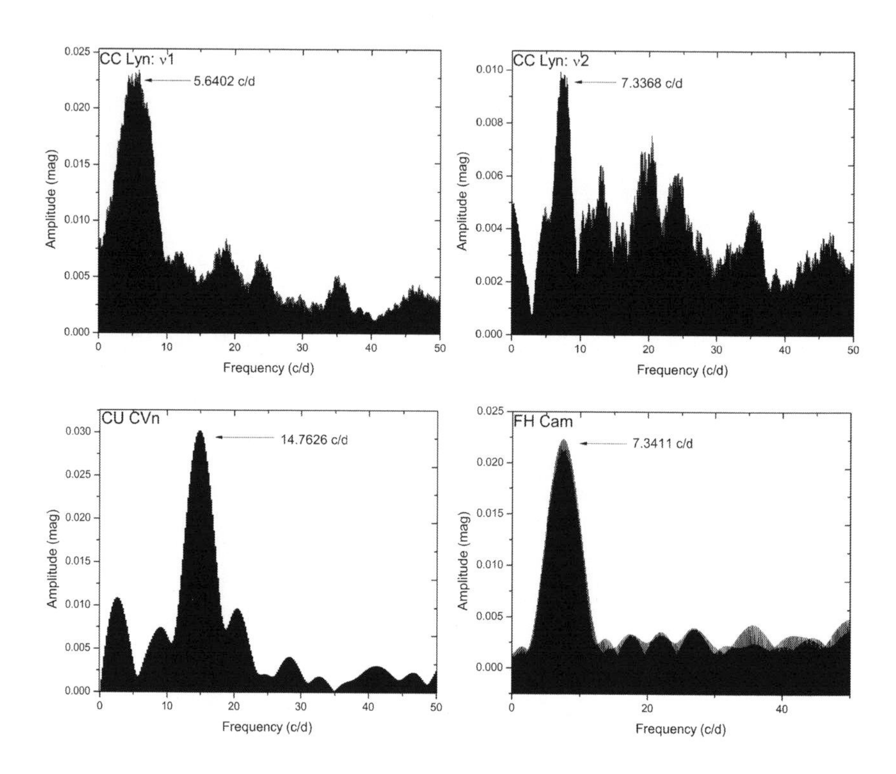

Figure 5: The amplitude spectra of CC Lyn, CU CVn, & FH Cam.

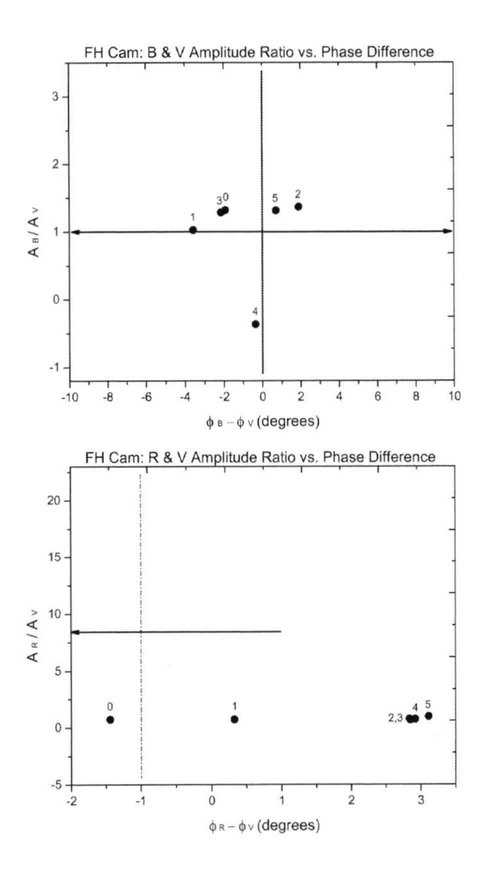

Figure 6: The Amplitude ratios vs. Phase differences for FH Cam are shown, for the B-V and R-V band photometry. The modal orders are indicated by the number print above a given data point. It can be seen that FH Cam must have a radial pulsation modal order, $\ell \leq 1$. The error values are indicated by the size of the dot used to indicate the appropriate modal order. The dotted vertical axis in the R-V diagram indicates that the vertical axis has been displaced from intersecting the origin so it can appear in the plot (origin is at $\sim$ (-106, 8)). The arrow heads indicate the measured phase difference was larger than plotted in the indicated direction: B-V phase = (-160, 160), V-R phase = (-215, 1) degrees.

5. Conclusion

We clearly see that all stars analyzed have the correct spectral type to be included in the instability strip for δ Scuti stars (Breger 2000), and have shown pulsations with $\nu > 5$ c/d. These results securely place FH Cam, CU CVn, and CC Lyn in the category of δ Scuti pulsators.

FH Cam is shown to have only one detectable pulsation, which is found to have $\ell \leq 1$. There is also a decrease in amplitude of the light curves as the filter's pass-band is centered at longer wavelengths, further supporting the claim that FH Cam is a δ Scuti pulsator. The variability in the systemic radial velocity also indicates FH Cam is part of a multiple system.

CU CVn is observed to have variability in both the photometric and spectroscopic data sets, implying we have detected a pulsation, with a frequency of 14.7626 ± 0.0250 c/d. Hence, the observed high frequency in conjunction with the spectral type being F0-2V confirms that CU CVn is a δ Scuti star. There was no noticeable variability of $RV_{\circ}$ for this object, but a longer baseline of data will be needed to rule out binarity.

We have found CC Lyn to have two detectable pulsations. The ratio of these two periods might indicate the fundamental and first overtone radial pulsations, but it is not quite definite. The change in systemic radial velocity confirms the binarity of CC Lyn.

Acknowledgments. The authors would like to express their appreciation to Prof. Slavek M. Rucinski for his support during the initiation of the project, and his guidance and invaluable discussions throughout all stages of the present paper. We also would like to express our gratitude to the telescope operators at the David Dunlap Observatory, Heidi DeBond, Wen Lu, and Jim Thompson, for their assistance during the observing runs and to Prof. Marshall L. McCall for his constructive advice. We finally express our gratitude to the anonymous referee for his insightful advice and comments regarding our work.

References

Berry, R., & Burnell, J., The Handbook of Astronomical Image Processing, Willmann-Bell, Richmond (Virginia), 2000

Breger, M. 2000, ASPC, 210, 3B

Grenier, S., Baylac, M.O., Rolland, L., et al. 1999, A&AS, 137, 451G

Kurtz, D. W. 2006, CoAst, 147, 6

Lenz, P. & Breger, M. 2005, CoAst, 146, 53

Perryman, M. A. C., Lindegren, L., Kovalevsky, J., et al. 1997, A&A, 323, 49

Pych, W. 2003, PASP, submitted (arXiv:astro-ph/0311290v1)

Pribulla, T., Rucinski, S. M., Blake, R. M., et al. 2009, AJ, 137, 3655

Rucinski, S. M. 1992, AJ, 104, 1968

Rucinski, S. M. 2002, PASP, 114 , 1124R

Rucinski, S., Lu, W. 1999, JRASC, 93, 186R

Vidal-Sáinz, J., Gomez-Forrellad, J. M., García-Melendo, E., Wils, P., & Lampens, P. 2002, IBVS, 5331, 1

Wilson, R.E. 1953,General Catalogue of Stellar Radial Velocities, Carnegie Institution of Washington Publ. 601, Washington D.C

Zima, W. 2008, CoAst, 155, 17

Comm. in Asteroseismology
Volume 161, June 2010

A comparison of Bayesian and Fourier methods for frequency determination in asteroseismology

T. R. White[1], B. J. Brewer[1,2], T. R. Bedding[1], D. Stello[1] and H. Kjeldsen[3]

[1] Sydney Institute for Astronomy (SIfA), School of Physics, University of Sydney, NSW 2006, Australia
[2] Department of Physics, University of California, Santa Barbara, CA 93106-9530, USA
[3] Danish AsteroSeismology Centre (DASC), Department of Physics and Astronomy, Aarhus University, DK-8000 Aarhus C, Denmark

Abstract

Bayesian methods are becoming more widely used in asteroseismic analysis. In particular, they are being used to determine oscillation frequencies, which are also commonly found by Fourier analysis. It is important to establish whether the Bayesian methods provide an improvement on Fourier methods. We compare, using simulated data, the standard iterative sine-wave fitting method against a Markov Chain Monte Carlo (MCMC) code that has been introduced to infer purely the frequencies of oscillation modes (Brewer et al. 2007). A uniform prior probability distribution function is used for the MCMC method. We find the methods do equally well at determining the correct oscillation frequencies, although the Bayesian method is able to highlight the possibility of a misidentification due to aliasing, which can be useful. In general, we suggest that the least computationally intensive method is preferable.

Accepted: June 7, 2010

1. Introduction

Bayesian methods are increasingly being used for asteroseismic analysis. Most effort has been directed at extracting mode parameters by fitting to the Fourier power spectrum (e.g. Appourchaux 2008; Benomar 2008; Benomar et al. 2009; Gaulme et al. 2009), but there have also been applications that involved fitting directly to the time series (Brewer et al. 2007; Brewer & Stello 2009). It is the latter approach, which can be thought of as an alternative to calculating the power spectrum, that is the subject of this paper. Determining the frequencies at which stars oscillate is fundamental to asteroseismology. The first step in doing so is generally to calculate the power spectrum of the time series, in which the frequencies of oscillation will appear as peaks. However, this is complicated by noise and aliasing so that it is not always immediately obvious which peaks are real. In this paper, we use simulated data to determine if a Bayesian Markov Chain Monte Carlo code is more effective at determining real frequencies than a standard iterative sine-wave fitting code.

1.1. Iterative sine-wave fitting

To systematically extract the peaks that are most likely to be real, an iterative algorithm (Roberts et al. 1987) is commonly used (e.g. Carrier & Bourban 2003; Kjeldsen et al. 2005; Bedding et al. 2007). This is referred to as iterative sine-wave fitting (also called CLEAN). In this algorithm, the discrete Fourier transform of the time series is calculated, generating the amplitude spectrum. The highest peak in the amplitude spectrum is identified and the corresponding sinusoid is then subtracted from the time series. The discrete Fourier transform of the residual time series is then calculated, and the process is repeated until the amplitude of the highest peak is below a chosen threshold. The result is a list of frequencies, amplitudes and phases that account for most of the variations in the time series.

However, this technique may run into difficulties. It is possible for alias peaks (sidelobes) to be enhanced by noise to the point where they have a higher amplitude in the Fourier spectrum than the genuine peak. In this case, iterative sine-wave fitting programs will identify the alias and subtract it from the time series. The amplitude of the genuine peak in the Fourier spectrum will be diminished by this process to the point where it may not be found at all. This problem may be exacerbated if the separation between oscillation peaks is approximately an integer multiple of the typical period between gaps in the data (usually one cycle per day in ground-based data) since the alias peaks may reinforce. In addition to this, noise peaks may obscure the signal.

Other concerns extend from the nature of Fourier methods. Bretthorst (1988) has shown that the power spectrum is an optimal procedure only in the case where only one mode is present, or if multiple, they are well separated in frequency. In practice, particularly with solar-like oscillations, this is rarely the case, with multiple modes often closely spaced in frequency. A further consideration is that the oscillation modes will vary in time such that a single mode must be represented by multiple sinusoids in the Fourier series. An additional problem can arise from the nature of the algorithm itself: when a frequency is found, and its amplitude and phase fitted to the time series, the corresponding sinusoid is subtracted. However as these parameters cannot be exact, biases are introduced into the residual time series.

1.2. Bayesian Methods

The above concerns have led to the consideration of Bayesian methods for determining frequencies. Probability theory may be used as a mathematical model of our belief in the plausibility of various hypotheses. Our knowledge of a set of parameters, θ, given prior information and assumptions, I, is represented by the *prior probability distribution*, $p(\theta|I)$. Given new data, D, *Bayes' theorem*,

$$p(\theta|D, I) \propto p(\theta|I)p(D|\theta, I), \tag{1}$$

tells us about our new state of knowledge, the *posterior distribution*. The distribution $p(D|\theta, I)$, the probability distribution for the data given the parameters, is a measure of how well the data are predicted by the model. In the case where the data are known and fixed, $p(D|\theta, I)$ becomes dependent on θ only and is called the *likelihood function*. How much we believe our model depends on both how well we originally believed it (the prior distribution), and on how well it predicts the new data. It is important to note that probabilities are always conditional on the underlying background information and/or assumptions I, even when these are not explicitly stated. Bayes' theorem provides a means of finding the most probable model that could produce the observed data.

A method using this probabilistic reasoning has been developed by Brewer et al. (2007) and applied to the subgiant stars ν Indi (Bedding et al. 2006; Carrier et al. 2007) and β Hydri (Bedding et al. 2007). This method utilises a version of the Metropolis-Hastings algorithm (Neal 1993), itself a Markov Chain Monte Carlo (MCMC) algorithm (Gregory 2005), to determine the most likely set of frequencies (represented by θ above) that could give rise to observed data. Essentially, the code samples the θ parameter space – the number of frequencies and their values – through a random walk with more time being spent in regions of higher probability. At each iteration of the code, the current state is randomly perturbed. If the perturbation results in a higher posterior probability then it is accepted as the new state, otherwise the perturbed state is accepted with a probability proportional to the ratio of the new posterior probability density to the old.

The method of Brewer et al. (2007) analyses the observed time series directly. To do this it finds the most likely sinusoids that fit the time series, assuming that the time series is composed of a small number of sinusoidal signals (one per mode) and Gaussian noise. This assumption is not generally valid due to the stochastic nature of excitation and damping of the oscillations.

Some methods that take this in to account are able to infer frequencies and line widths by fitting to the power spectrum (e.g. Appourchaux 2008). However, the presence of gaps in the data, and the stochastic nature of oscillations results in the possibility that information will be lost in the power spectrum (Bretthorst 1988). It was this potential loss of information from using the power spectrum that was one of the reasons for considering Bayesian methods in the first place. The convenience of the power spectrum cannot be doubted and for data with good coverage and long mode lifetimes, the loss of information is negligible. Nevertheless it would be ideal to have a method that both analyses the time series directly and takes into account that the oscillations are not purely sinusoidal. Such a method has been developed (Brewer & Stello 2009), but is unfortunately computationally intensive and is currently only practical for short time series (fewer than ~ 1500 points). For this reason, it is the method of Brewer et al. (2007) we use in our comparison here with Fourier methods.

A major (although potentially dangerous) feature of Bayesian inference is the ability to incorporate extra knowledge in determining which parameters are more likely by choosing a descriptive prior distribution. For some stellar oscillations, namely high-overtone, low-degree acoustic oscillations in spherically symmetric stars, modes are usually expected to follow the asymptotic relation,

$$\nu_{n,l} = \Delta\nu(n + \frac{1}{2}l + \epsilon) - l(l+1)D_0, \tag{2}$$

where $\Delta\nu$ is called the large separation and depends on the sound travel time across the whole star, D_0 is sensitive to the sound speed near the core and ϵ is sensitive to the surface layers (Tassoul 1980, 1990). With this in mind, a prior distribution could be chosen in which a regular spacing of modes is anticipated. The MCMC code would then favour finding peaks with a regular spacing over finding peaks that do not fit this pattern. This was a feature of the code when it was introduced by Brewer et al. (2007) and tested on simulated data (which did have a constant separation between modes) and when it was used on real observations of ν Indi by Bedding et al. (2006) to find the large separation. The value inferred by the Bayesian method for the large separation, $\Delta\nu = 24.25 \pm 0.25\mu$Hz, agreed well with the value obtained from the peak of the autocorrelation function of the power spectrum, a Fourier method.

As mentioned, this use of a descriptive prior distribution can be dangerous. Although the modes of oscillation will be separated in frequency by approximately equal amounts, in general there will be a departure from the asymptotic relation. The large separation itself may be frequency-dependent. By using a prior that is too prescriptive, this may be missed by the Bayesian method. To avoid this possibility, the method described by Brewer et al. (2007) incorporated a uniform component and a regular pattern of Gaussian peaks, as opposed to delta functions. However, this may still be too prescriptive. A further analysis of the data on ν Indi showed that the large separation was indeed a function of frequency, with the average large separation of $\Delta\nu = 25.14 \pm 0.14\mu$Hz larger than first realised (Carrier et al. 2007).

Further use of this method was made by Bedding et al. (2007) on β Hydri. They used the Bayesian method as a supplement to traditional Fourier methods to determine individual frequencies, although no assumptions about the frequency distribution were made. To avoid detecting noise peaks in the Fourier analysis, only the peaks with a signal-to-noise ratio (S/N) ≥ 4 were included. Comparing the results with the Bayesian method, one extra peak was found by the Bayesian method, lying precisely on the $l = 1$ ridge of the échelle diagram. The same peak was found by the Fourier iterative sine-wave fitting with S/N $= 3.0$.

These studies of ν Indi and β Hydri have obtained similar results using both Bayesian and traditional Fourier method. The question arises as to which method, if either, is superior. Brewer et al. (2007) suggested that the Bayesian method was less susceptible to aliasing and, by effectively fitting multiple sinusoids simultaneously, was less susceptible to any issues caused by the subtraction of modes from the time series. On the other hand, the Bayesian method is more computationally intensive for large data sets, and it has been suggested that its advantage over the traditional Fourier methods is with shorter, noisier, and more incomplete data sets. Despite these presumed advantages, a detailed comparison of the methods has not yet been made. It is the purpose of this paper to do this comparison on simulated data for which the frequencies are known.

2. Programs being tested

Two different iterative sine-wave fitting programs were used, both written by one of us (HK). They differ in that one program re-adjusts the frequencies, amplitudes and phases of all previously extracted peaks at each iteration, in an attempt to improve the fit. However, there is a time penalty to this, and it may not necessarily lead to improved results because the process may be unstable. Hereafter, the version which does not recalculate parameters at each iteration is referred to as *fast* and the other as *slow*.

The program used to evaluate Bayesian methods of determining frequencies was written by one of us (BJB) and is a variation of the program introduced by Brewer et al. (2007), which implements a Markov Chain Monte Carlo (MCMC) algorithm. The principles of the original program have been discussed above. Here, we outline how our implementation of the code differs.

The most important difference is the choice of the prior probability distribution. Our choice here can have a significant impact on the results obtained.

Brewer et al. (2007) used a prior distribution which took into account that the frequencies of low-degree $p-$mode oscillations are expected to be approximately given by Equation 2. This is not a valid assumption if a mode exhibits mixed $p-$ and $g-$mode behaviour, as in evolved stars, or if the large separation, $\Delta\nu$, is not constant with frequency. Setting this prior will have a significant impact on the posterior distribution, potentially placing unreasonably high confidence in low signal-to-noise peaks that fit a regular spacing, and down-weighting real peaks that do not. For our implementation of this method, we have chosen only to use a uniform prior that does not anticipate the separation of peaks to avoid the possibility of the code finding a regular spacing that may not be there. The uniform prior is given by

$$g = \frac{1}{\nu_{\mathrm{max}} - \nu_{\mathrm{min}}}, \tag{3}$$

where $(\nu_{\mathrm{min}}, \nu_{\mathrm{max}})$ is the range of frequencies over which we attempt to find oscillations.

Since Brewer et al. (2007) used a sharply peaked prior, it was possible for the program to become stuck in a local minimum and not adequately sample the parameter space. To overcome this problem parallel tempering was used with respect to the prior. In our case we do not use parallel tempering since the uniform prior is without the sharp peaks that could cause the chains to become stuck in local minima.

Will this change to the prior have much impact on the number of frequencies used in the models at each iteration? That is, how effective was the use of a prior that expected a regular spacing of peaks at suppressing peaks that did not meet this criteria? Figure 3 of Brewer et al. (2007) shows the output of a MCMC code run on simulated data that contained 17 frequencies. The code infers that there are at least 16 frequencies, and possibly up to about 20, with 16 the most probable solution. A typical output of the MCMC code used in this paper is shown in Figure 1 for the simulated data discussed in Section 3. After a short initial burn-in period, the distribution of the number of frequencies settles into the posterior distribution. The number of frequencies fluctuates between approximately 50 and 150, averaging around 90. The actual number used was 61. Although our simulated data is different to that of Brewer et al. (2007), it does appear that the uniform prior results in additional frequencies being accepted than with the descriptive, regularly peaked prior, as could be expected. Since we do expect there to be significant departures from an equal spacing in some stars, we are prepared to accept the increased chance of detecting noise peaks and aliases between equally spaced frequencies for this comparison with Fourier methods.

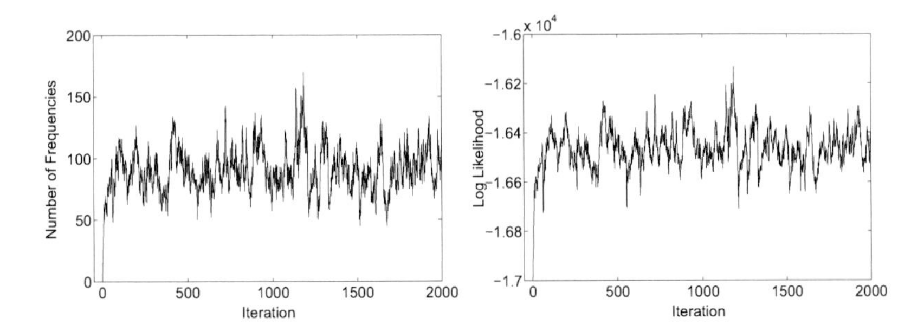

Figure 1: Results of a MCMC run with a uniform prior. *Left* Number of frequencies fitted as a function of iteration. *Right* Log likelihood of the proposed model as a function of iteration.

The original code, in comparison to traditional methods, had the deficiency of not returning the amplitudes of the modes, but only their relative probabilities. This has since been rectified, with the Markov Chain now sampling the distribution of both frequencies and amplitudes at each iteration.

3. Tests with coherent oscillations

Our first test of the algorithms was the simplest case possible, in which the oscillations are pure sinusoids and are well-separated in frequency. While no stellar oscillation is entirely coherent, there exist many cases where the mode lifetime is long enough for this to be a good approximation. For this test it was important to probe a variety of signal-to-noise (S/N) ratios to gauge the performances of the programs at different levels. S/N ratios from 2 and 5 were probed in this test.

The time series was generated as a sum of 61 sinusoids with frequencies separated by 60 μHz, ranging from 300 μHz to 3900 μHz. Phases were chosen at random. The generated signal was sampled every 100 seconds for a total of 9 days. Gaps are common in astronomical datasets, primarily due to the usual restriction of observations to be taken at night, which causes the aliasing that is so problematic to frequency analysis. To simulate this, only the first 40 per cent of a 'day' of simulated observations was retained as part of the time series. The resulting time series had 3114 data points. Random Gaussian noise with a standard deviation of 3.09 ms^{-1} was added to the data, which was calculated to provide an average noise in the amplitude spectrum of 0.10 ms^{-1} (from equations A1 and A2 of Kjeldsen & Bedding 1995). The amplitudes chosen for the sinusoids varied linearly from 0.2 ms^{-1} at 300 μHz up to 0.5 m $^{-1}$ at

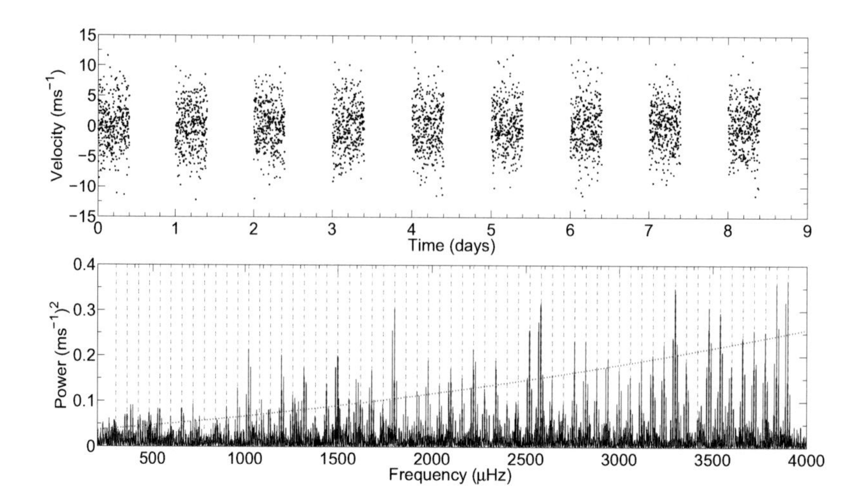

Figure 2: *Top* Simulated time series with coherent oscillations ranging in S/N from 2 to 5. *Bottom* Corresponding power spectrum. Input frequencies are indicated by the vertical blue dashed lines. Input power as a function of frequency is indicated by the red dotted line. Sidelobes at $\pm 11.6\mu$Hz due to daily gaps are clearly visible. Colour available on online version.

3900 μHz, to give the desired range of S/N ratios. It should be noted that, due to the effects of noise, the measured S/N of these peaks will differ from the input values, with some peaks being suppressed and others enhanced. The final time series is shown in Figure 2, together with its power spectrum. It is apparent that many more peaks are present in the power spectrum than the 61 input sinusoids marked by dashed lines, with some due to aliasing occurring at $\pm 11.6\mu$Hz (± 1 cycle/day) from the input peaks and others due to noise.

3.1. Results

The simulated time series was analysed with both versions of the iterative sine-wave fitting program, and with the Bayesian MCMC code. The peaks extracted by the fast iterative sine-wave fitting program and the probability spectrum output by the MCMC code are shown in Figure 3. It is immediately apparent from these graphs that both methods successfully identified the modes with the highest S/N and that they agree in general. Although the relative heights of peaks are different between the two spectra in Figure 3, the modes with the highest S/N generally correspond to those with the greatest posterior probability.

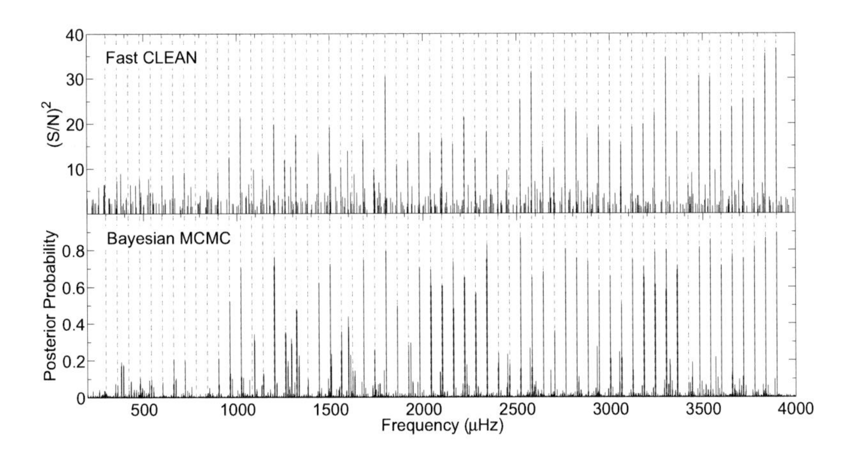

Figure 3: *Top* Peaks extracted by the fast iterative sine-wave fitting program. *Bottom* Probability spectrum from the Bayesian MCMC program. Blue dashed lines indicate the input frequencies.

To further see this relation between probability and S/N, Figure 4 compares the probability and S/N of matched peaks between the Bayesian and Fourier methods. Between a S/N of 2 and 4 there is a roughly linear relationship between S/N and posterior probability. Either side of this region the probability saturates, resulting in a flattening of the graph. From this figure it is apparent that peaks can generally be believed (open circles) if the S/N is above $\sim$3.5 or the posterior probability is above $\sim$0.4.

If a peak was identified by one program only, whether real, alias or noise, it appears on this graph on the vertical line of points at a S/N of $\sim$1.1 or the horizontal line of points at a posterior probability of $\sim$0.01. If the iterative sine-wave fitting program did not find a peak above the S/N$\sim$1.1 threshold it is either due to the peak having an inherently low S/N, or to the peak being diminished when a sine-wave was removed from the time series in previous iterations. This second explanation is responsible for the number of alias peaks in this category, which would be effectively removed in the Fourier case if the correct peak were identified first. However, if it happened that the fast Fourier program mistakenly identified the alias of a given peak, then it would not later detect the correct peak. The slow Fourier program would have a chance to change its identification, though in practice, this turned out to be rare. We found that the Bayesian program is more likely to identify both, and so may be of use as an alert to the possibility that a peak has been misidentified with

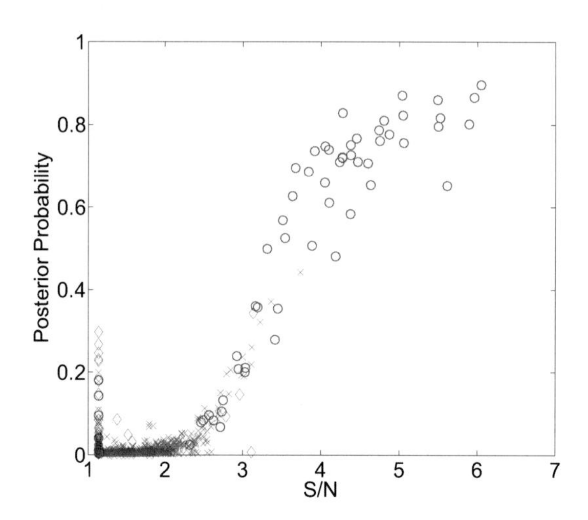

Figure 4: Posterior probability from Bayesian MCMC versus S/N from fast iterative sine-wave fitting program. Blue circles correspond to true input frequencies, magenta diamonds are aliases at $\pm 11.6\,\mu$Hz, and red crosses are noise peaks.

its alias. Comparing the échelle diagrams of the fast iterative sine-wave fitting program with the MCMC Bayesian code in Figure 5, we see that they effectively do equally well at identifying the correct peaks.

Despite the much longer time required to run the slow version of the iterative sine-wave fitting program, there did not appear to be any significant difference to the fast version, as can be seen in Figure 5. In general, the slow version is slightly more accurate in determining the frequencies, although in one case, the slow version has wrongly chosen the first and second aliases of the correct peak when the fast version did not. Stello et al. (2006) similarly found that the difference between the two was minor. They found that the fast version identified 1% false peaks compared to a perfect result by the slow version for simulated data from 100 time series with 17 equally-spaced frequencies with no noise, extracting 10 frequencies per time series. The scatter of the output frequencies relative to the input was roughly equal to the frequency resolution for the slow version, and about twice that size for the fast version. For non-coherent oscillations, the differences were diminished. The slow version takes a factor of $1.5(N+1)$ longer than the fast version, where N is the number of frequencies to be extracted (Stello et al. 2006) and so, for a large number of frequencies, is prohibitively slow.

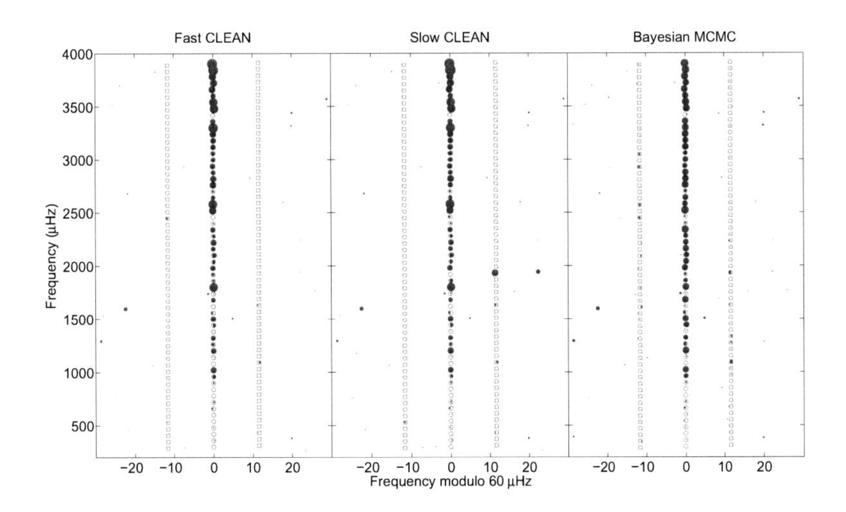

Figure 5: *Left* Échelle diagram from fast iterative sine-wave fitting program. *Centre* Échelle diagram from slow iterative sine-wave fitting program. *Right* Échelle diagram from Bayesian MCMC. The blue open circles indicate the location of real input peaks, with the green open squares indicating their aliases. Red filled circles indicate the peaks found by each program, with the size indicating the relative strength of each peak. The sizes have been scaled, so that peaks that follow the linear trend in Figure 4 will have approximately the same size in each diagram.

4. Tests with problematic aliasing

The next test investigates situations where aliasing is particularly problematic, that is, when the frequency separation between modes is close to an integer multiple of one cycle/day. The first side-lobes of adjacent peaks will coincide when the separation is exactly 2 cycles/day. A chirp was introduced to the input frequencies so that the frequency separation increased from 1.8 cycles/day up to 2.2 cycles/day, at a rate of 0.1 cycles/day/order. The input S/N of each peak was fixed to 3.5. As previously mentioned, the effects of noise will modulate this in the output time series.

The window function and noise was the same as in the previous test (Section 3). The power spectrum of this time series is shown in Figure 6. The aliases are clearly apparent, approximately halfway between the real input peaks (dashed lines). Many of the sidelobes are higher than real peaks.

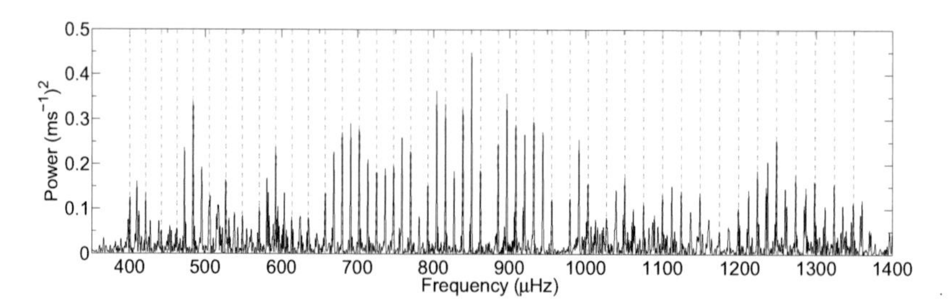

Figure 6: Power spectrum of simulated time series with problematic aliasing. Input frequencies are indicated by the blue dashed lines.

4.1. Results

Both programs performed quite well when the separation was significantly different from two cycles/day. When the aliases of adjacent modes coincide (in the middle of Figure 6), both failed to detect the correct frequencies. Matching the frequencies extracted by each program and plotting posterior probabilities against S/N in Figure 7, we see the same general trend as shown in Figure 4, although there is clearly more scatter. This scatter is due to the Bayesian MCMC code alternatively sampling both the real and alias peaks, and down- weighting their relative probabilities, whereas the iterative sine-wave fitting code will identify the highest peak at its S/N, real or alias, and miss the other altogether. This explanation is further borne out by the échelle diagrams, shown in Figure 8. The Bayesian method does have an advantage in drawing attention to the possibility of confusion, but is not any better at determining which is real.

5. Discussion and Conclusions

The Bayesian program and the traditional Fourier methods do equally well at identifying the correct frequency of stellar oscillations. The Bayesian method is effectively attempting to fit multiple sinusoids to the time series, whereas the iterative sine-wave fitting program finds the highest peaks, one at a time, from the Fourier amplitude spectrum. This often leads the Bayesian program to sample both the real peak and its aliases, whereas the Fourier program will identify only the highest of these. Without a more descriptive prior probability, it is not possible for the Bayesian program to avoid identifying alias or noise peaks that have high S/N. However, a more detailed prior might cause the results to be dominated by what was expected and not by what is actually present. It may be the case that the frequencies that are unexpected prove to be the most interesting.

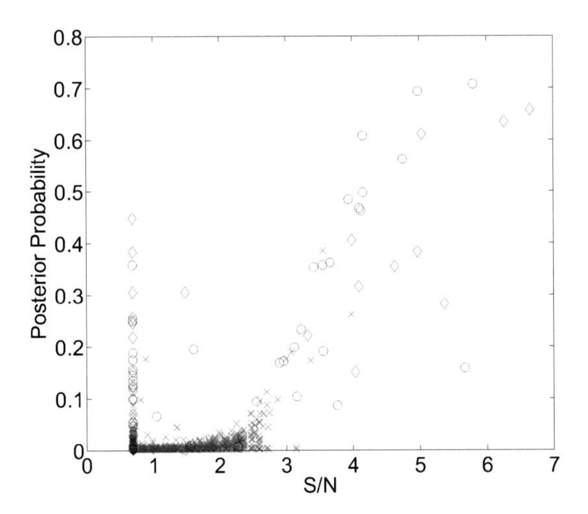

Figure 7: Same as for Figure 4 but for a time series with problematic aliasing.

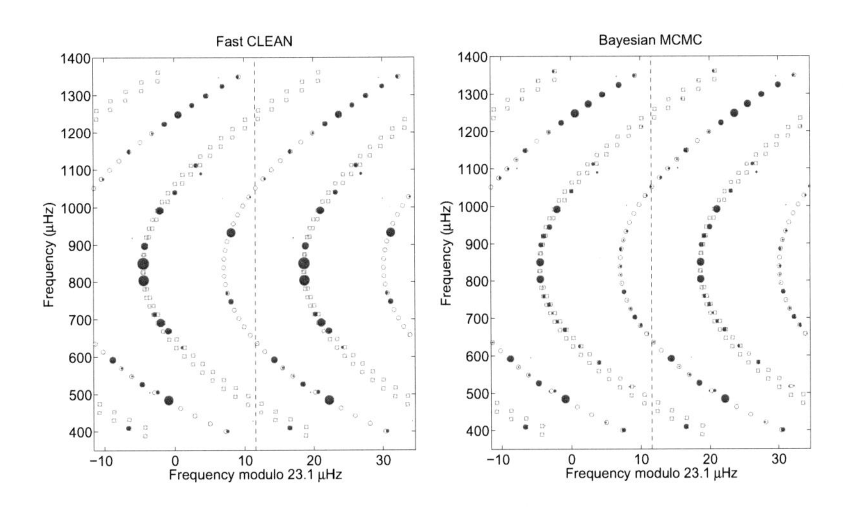

Figure 8: Échelle diagrams for a time series with problematic aliasing. *Left* Fast iterative sine-wave fitting, *right* Bayesian MCMC. Symbols same as for Figure 5. Note that the échelle diagrams have been plotted twice so that a ridge of input frequencies is continuous, with the dashed line indicating the half-way join. The ridge of input frequencies, and their aliases are curved due to the varying separation of the input series.

When aliases are strong, the Bayesian method does have an advantage in highlighting the possibility of the confusion because it should detect both aliases and real peaks. In general however, we have found no advantage of the Bayesian method over the traditional Fourier methods. It is therefore recommended that the least computationally intensive program be used. The fast Fourier program was found to be fastest, provided there were not many peaks to be extracted, and there are a large number data points ($>\sim$ 10000). If a large number of peaks is required, and there are not too many data points, then the Bayesian method takes a comparable time. The slow Fourier program is prohibitively slow for determining any more than a few frequencies.

While we have shown that there is no major advantage in using the Bayesian approach discussed here, it is clear that Bayesian methods will continue to be used for fitting to the power spectrum once it has been calculated using traditional Fourier methods.

Acknowledgments. We acknowledge support from the Australian Research Council. TRW is supported by an Australian Postgraduate Award, a University of Sydney Merit Award and a Denison Merit Award.

References

Appourchaux, T. 2008, Astronomische Nachrichten, 329, 485

Bedding, T. R., Butler, R. P., Carrier, F., et al. 2006, ApJ, 647, 558

Bedding, T. R., Kjeldsen, H., Arentoft, T., et al. 2007, ApJ, 663, 1315

Benomar, O. 2008, Communications in Asteroseismology, 157, 98

Benomar, O., Appourchaux, T., & Baudin, F. 2009, A&A, 506, 15

Bretthorst, G. L. 1988, Lecture Notes in Statistics, Vol. 48, Bayesian Spectrum Analysis and Parameter Estimation (Springer-Verlag, New York)

Brewer, B. J., Bedding, T. R., Kjeldsen, H., & Stello, D. 2007, ApJ, 654, 551

Brewer, B. J. & Stello, D. 2009, MNRAS, 395, 2226

Carrier, F. & Bourban, G. 2003, A&A, 406, L23

Carrier, F., Kjeldsen, H., Bedding, T. R., et al. 2007, A&A, 470, 1059

Gaulme, P., Appourchaux, T., & Boumier, P. 2009, A&A, 506, 7

Gregory, P. C. 2005, Bayesian Logical Data Analysis for the Physical Sciences: A Comparative Approach with 'Mathematica' Support (Cambridge University Press)

Kjeldsen, H. & Bedding, T. R. 1995, A&A, 293, 87

Kjeldsen, H., Bedding, T. R., Butler, R. P., et al. 2005, ApJ, 635, 1281

Neal, R. M. 1993, Probabilistic Inference Using Markov Chain Monte Carlo Methods, Technical Report CRG-TR-93-1, Dept. of Computer Science, University of Toronto, available at http://www.cs.toronto.edu/~radford/

Roberts, D. H., Lehar, J., & Dreher, J. W. 1987, AJ, 93, 968

Stello, D., Kjeldsen, H., Bedding, T. R., & Buzasi, D. 2006, A&A, 448, 709

Tassoul, M. 1980, ApJS, 43, 469

Tassoul, M. 1990, ApJ, 358, 313

Comm. in Asteroseismology
Volume 161, June 2010

A new eclipsing binary system with a pulsating component detected by CoRoT*

K. Sokolovsky[1,2], C. Maceroni[3], M. Hareter[4], C. Damiani[3], L. Balaguer-Núñez[5], and I. Ribas[6]

[1] Max-Planck-Institute für Radioastronomie, Auf dem Hügel 69, D-53121 Bonn, Germany
[2] Astro Space Center of Lebedev Phys. Inst., Profsoyuznaya 84/32, 117997 Moscow, Russia
[3] INAF - Osservatorio Astronomico di Roma, via Frascati 33, Monteporzio C., Italy
[4] Institut für Astronomie, Universität Wien, Türkenschanzstraße 17, 1180 Vienna, Austria
[5] Departament d'Astronomia i Meteorologia-ICC-IEEC, Universitat de Barcelona, Av. Diagonal, 647, 08028 Barcelona, Spain
[6] Institut de Ciències de l'Espai (CSIC-IEEC), Campus UAB, Facultat de Ciències, Torre C5, parell, 2a pl., E-08193 Bellaterra, Spain

Abstract

We report the discovery of CoRoT 102980178 ($\alpha = 06^{h}50^{m}12\overset{s}{.}10$, $\delta = -02°41'$ $21\overset{''}{.}8$, J2000) an Algol–type eclipsing binary system with a pulsating component (oEA). It was identified using a publicly available 55 day long monochromatic light curve from the CoRoT initial run dataset (exoplanet field). Eleven consecutive $1\overset{m}{.}26$ deep total primary and the equal number of $0\overset{m}{.}25$ deep secondary eclipses (at phase 0.50) were observed. The following light elements for the primary eclipse were derived: $HJD_{\rm MinI} = 2454139.0680 + 5\overset{d}{.}0548 \times E$. The light curve modeling leads to a semidetached configuration with the photometric mass ratio $q = 0.2$ and orbital inclination $i = 85°$. The out-of-eclipse light curve shows ellipsoidal variability and positive O'Connell effect as well as clear $0\overset{m}{.}01$ pulsations with the dominating frequency of 2.75 c/d. The pulsations disappear during the primary eclipses, which indicates the primary (more massive)

*The CoRoT space mission was developed and is operated by the French space agency CNES, with participation of ESA's RSSD and Science Programs, Austria, Belgium, Brazil, Germany and Spain.

component to be the pulsating star. Careful frequency analysis reveals the second independent pulsation frequency of 0.21 c/d and numerous combinations of these frequencies with the binary orbital frequency and its harmonics. On the basis of the CoRoT light curve and ground-based multicolor photometry, we favor the classification of the pulsating component as a γ Doradus type variable, however, the classification as an SPB star cannot be excluded.

Accepted: June 25, 2010

Individual Objects: CoRoT 102980178

1. Introduction

Pulsating stars in eclipsing binary systems are objects of considerable astrophysical interest. It is well known that in double–line binaries masses, radii and luminosities of the components can directly be determined allowing us to constrain pulsation models. Possible influence of one component of a binary system on pulsations of the other component (through mass transfer or tidal interaction) is an interesting and still largely unexplored topic. A case of tidal excitation of γ Doradus type pulsations was reported by Handler et al. 2002 (see also references to some theoretical discussions therein). In binary systems with one oscillating component the change of radius can be directly observed during the pulsation cycle by determining periodic changes in the times of the first to fourth contact. Bíró & Nuspl (2005) suggest that the pulsation mode can be identified by using the secondary component as a spatial filter. Altogether, the identification of pulsating stars of different types in eclipsing systems is important.

The combination of pulsations and eclipses is often found among the evolved stars like symbiotic binaries (e.g. Chochol & Pribulla 2000) and cataclysmic variables (e.g. Araujo-Betancor et al. 2005). Four Type I or II Cepheids in eclipsing binaries have recently been identified (see Antipin et al. 2007 and references therein), the confirmation of one more possible object of this type is still pending (Khruslov 2008). The discovery of a red giant with solar-like oscillations in a long-period eclipsing binary system was recently announced by Hekker et al. (2010).

Semidetached eclipsing binaries with a pulsating primary component which is close to the main sequence were dubbed "oscillating Algols" (oEA, Mkrtichian et al. 2002). About twenty such systems are known to date (Mkrtichian et al. 2007), most of them containing δ Scuti type components. The first eclipsing binary containing a γ Doradus type pulsating star (VZ CVn) was reported by Ibanoğlu et al. (2007). Recently, Damiani et al. (2010) and Maceroni et al. (2010) identified two γ Doradus candidates in eclipsing binaries using CoRoT photometry (CoRoT 102931335 and CoRoT 102918586).

CoRoT (Convection, Rotation and planetary Transit, Fridlund et al, 2006) is a space experiment devoted to study the convection and rotation of stars and to detect planetary transits. CoRoT data become publicly available one year after the release to the Co–Is of the mission from the CoRoT archive: `http://idoc-corot.ias.u-psud.fr/`

In this paper we announce the discovery of a new oEA system identified using publicly available data from the CoRoT initial run (exoplanet field). An almost uninterrupted 55 day long monochromatic CoRoT light curve together with ground–based multicolor photometry allowed us to identify the pulsating component as a likely γ Dor type variable, however, the classification as a Slowly Pulsating B (SPB) star cannot be ruled out. The object is an interesting target for a spectroscopic follow–up.

2. Observational data

2.1. CoRoT photometry

CoRoT 102980178 (USNO-B1.0 0873-0161681, coordinates: $\alpha = 06^{h}50^{m}12\overset{s}{.}10$, $\delta = -02^{\circ}41'21\overset{''}{.}8$, J2000, Monet et al. 2003) was identified by us as an Algol type eclipsing binary with an oscillating component (oEA star) after the visual inspection of the CoRoT light curve. The star was also independently identified as an eclipsing binary by Debosscher et al. (2009) and Carpano et al. (2009). The CoRoT light curve covers almost 55 days from JD 2454138.1 to JD 2454192.8 with 512 sec time sampling. We have removed data points with large error–bars which were typically upward outliers caused by high–energy particles hitting the CCD detector inside the measurement aperture. Most of these events occur when the satellite crosses the South Atlantic Anomaly (SAA). Elimination of these data points introduces small gaps in the light curve, with some of these gaps occurring quasi–periodically (due to SAA crossing). However, the number of discarded data points is relatively small (1168 out of 9229, 12.6%). To characterize the effect of periodic data rejection on the frequency analysis we have constructed the spectral window plot presented on Fig. 1. The amplitudes of the alias frequencies are of the order of 10 % or below. We have also corrected the light curve for a small (0.211 mmag/day) downward trend which is probably a result of a gradual drift of the satellite pointing or some kind of instrumental decay. A detailed discussion of noise properties of the CoRoT data can be found in Aigrain et al. (2009) and Auvergne et al. (2009).

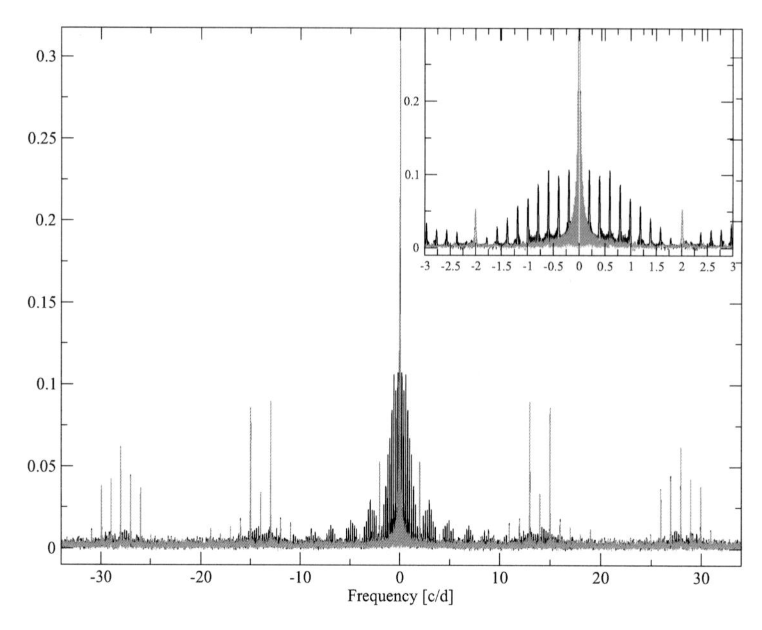

Figure 1: Spectral windows for the outlier-rejected data (light) and for the data excluding the primary eclipses (dark), which were used for frequency analysis. The inset shows a zoom-in to the lower frequencies.

2.2. Ground–based multicolor data

In the framework of ground support to the CoRoT mission, we obtained Strömgren *uvby* photometry of this object using the Wide Field Camera (WFC) on the 2.5 m Isaac Newton Telescope (INT) at Roque Muchachos Observatory, La Palma, Canary Islands. The observations were conducted on JD 2454170.391 which corresponds to the orbital phase 0.197 of the binary system. The object was also observed in Johnson B and V filters by Deleuil et al. 2006 using the same telescope and camera. Infrared JHK_s photometry of this object is available from the 2MASS catalog (Skrutskie et al. 2006). The 2MASS observations were conducted on JD 2451526.704 which corresponds to the orbital phase 0.191 (by a lucky coincidence this is very close to the phase of our Strömgren photometry). The great advantage of the 2MASS photometry is that it was obtained using a camera which utilizes a beam splitter producing truly simultaneous measurements in three filters. Therefore, 2MASS colors are not distorted by stellar variability.

All available color information is summarized in Table 1.

Table 1: Multicolor photometry of CoRoT 102980178

Parameter	Value	Error	Origin
$b-y=$	0.598	±0.032	INT/WFC
$y=$	15.723	±0.034	INT/WFC
$m_1=$	-0.042	±0.11	INT/WFC
$c_1=$	0.891	±0.086	INT/WFC
$B=$	16.58	±0.45	INT/WFC
$V=$	15.73	±0.19	INT/WFC
$J=$	13.411	±0.028	2MASS
$H=$	12.759	±0.023	2MASS
$K_s=$	12.528	±0.027	2MASS

2.3. Interstellar extinction estimation

In the absence of spectroscopic data it is important to constrain the influence of interstellar reddening, because information about intrinsic colors is necessary for an unambiguous classification of the pulsating star.

According to the standard tables by Schlegel et al. (1998) the *total* Galactic extinction in the direction of the object is large: $A_V = 3\overset{m}{.}860$, $E(B-V) = 1\overset{m}{.}164$. The Schlegel et al. estimation is based on direct observations of far–infrared emission of interstellar dust, material which causes absorption in visible and near–infrared bands. However, as it is noted by the authors, the extinction estimation for this position may be unreliable because of the low Galactic latitude ($b = -1\overset{\circ}{.}50$). There is an alternative way to estimate the Galactic extinction which is based on the relation between HI column density (N_{HI}) and extinction caused by dust: $N_{HI}/A_V = 1.79 \times 10^{21}$ cm^{-2} mag^{-1} (Predehl & Schmitt 1995). We estimate N_{HI} in the direction of the object using 21 cm radio observations from the Leiden/Argentine/Bonn Galactic HI Survey (Kalberla et al. 2005, see also `http://www.astro.uni-bonn.de/~webaiub/english/tools_labsurvey.php`). From these data we obtain $N_{HI} = 0.682 \times 10^{22}$ cm^{-2} corresponding to $A_V(HI) = 3\overset{m}{.}810$ which is in good agreement with the estimation obtained using the Schlegel et al. tables.

Unfortunately, the distance to the binary system is unknown, as well as the exact distribution of the absorbing material along the line of sight. The intrinsic color of the system lies somewhere between the observed color and the color corrected for the total Galactic reddening along the line of sight. However, even reliable upper limits on the Galactic reddening will be useful for the following discussion.

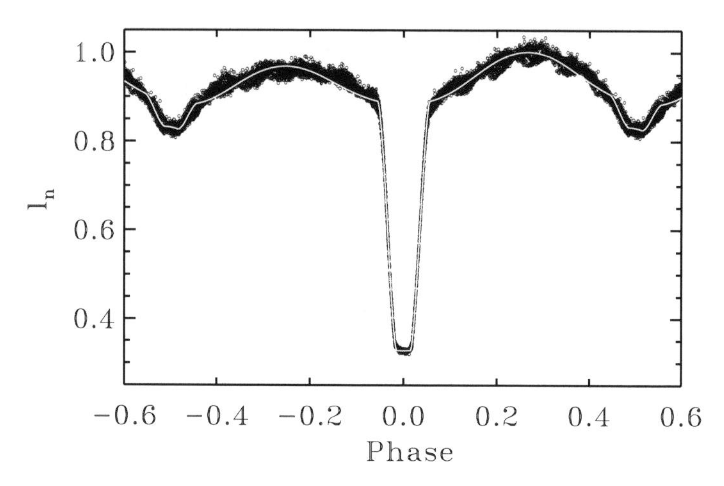

Figure 2: Detrended light curve of CoRoT 102980178 phased with light elements $HJD_{\mathrm{MinI}} = 2454139.0680 + 5\overset{d}{.}0548 \times E$. The solid line represents the F-primary model light curve.

3. Interpretation and Modeling

The light curve of CoRoT 102980178 shows eleven $1\overset{m}{.}26$ deep primary and the equal number of $0\overset{m}{.}25$ deep secondary eclipses (at phase 0.50). The nearly flat primary minimum bottom indicates that the eclipses are total. Ellipsoidal variability is evident in the out of eclipse light curve (Fig. 2). The maximum preceding the primary minimum is $0\overset{m}{.}04$ fainter than the maximum which follows it (positive O'Connell effect, see Davidge & Milone 1984 and Liu & Yang 2003 for a discussion of the effect). The period analysis with the Lafler & Kinman (1965) method leads to the following light elements:

$$\begin{array}{lrcrcl} HJD_{\mathrm{MinI}} = & 2454139.0680 & + & 5\overset{d}{.}0548 & \times & E \\ & \pm 0.0007 & & \pm 0\overset{d}{.}0170 & & \end{array}$$

Individual minima times were estimated using the Kwee & van Woerden (1956) method and combined to produce a single primary minimum epoch. Its 1σ uncertainty was estimated from the scatter of individual measurements on the Observed minus Calculated ($O - C$) plot produced with the above light elements. Period uncertainty was estimated following Schwarzenberg-Czerny (1991).

Superimposed on this classical Algol–type light curve are $0\overset{m}{.}01$ oscillations with the period of $0\overset{d}{.}36372$ (see Fig. 3 and the detailed discussion below). The oscillations are evident at all phases except during the primary minimum (Fig. 4).

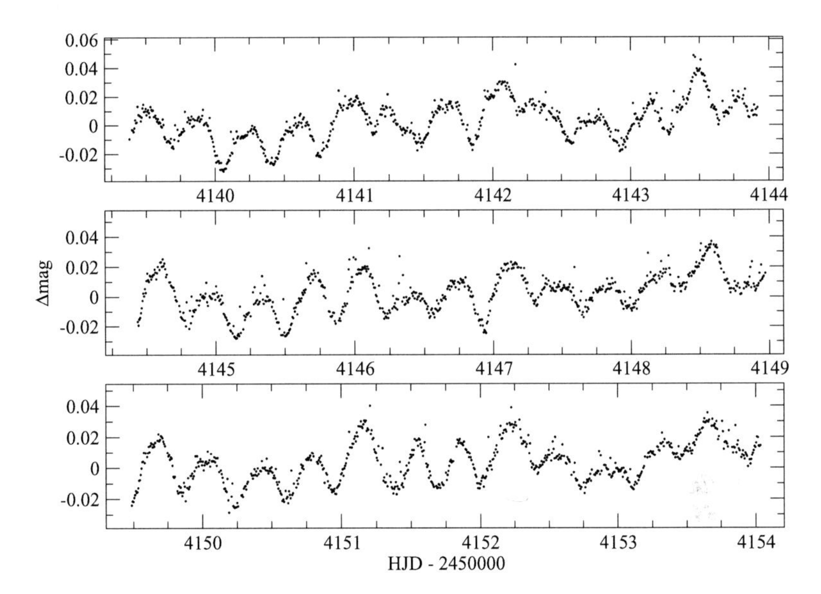

Figure 3: The residual light curve (F-primary model).

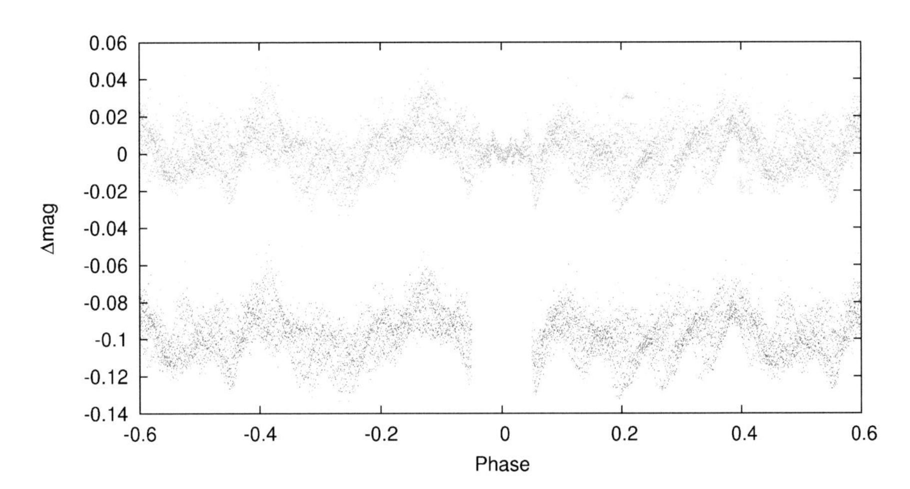

Figure 4: The residual light curve (F-primary model) folded with the binary orbital period. The lower curve represents the section of the residual light curve used for frequency analysis, it was shifted $0^{\mathrm{m}}.1$ along the vertical axis for visibility.

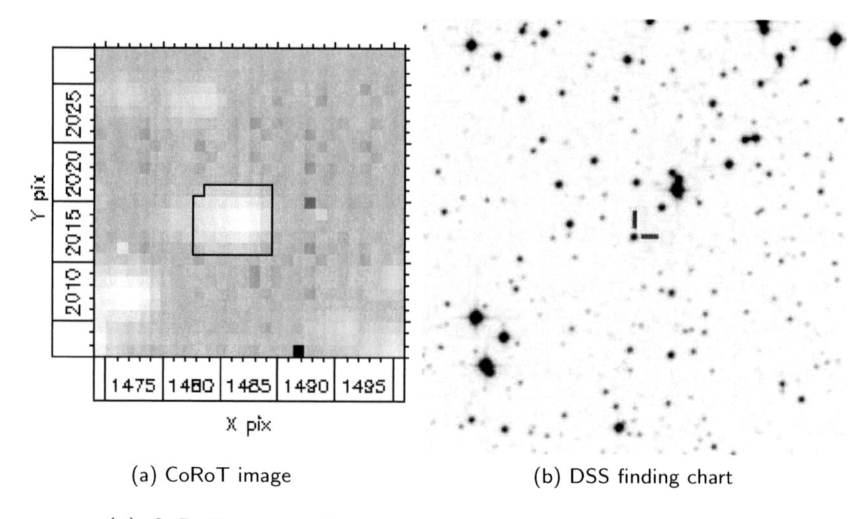

(a) CoRoT image (b) DSS finding chart

Figure 5: (a) CoRoT image of the variable star, the marked area is the measurement aperture. The image is in detector coordinates, east is up, north is to the left. (b) 5′ x 5′ DSS2 red image centered on CoRoT 102980178 (north is up, east is to the left).

3.1. Ruling out an optical blend

When analyzing a variable star showing two different types of variability (e.g. eclipses and pulsations) it is possible that the observed variability comes from two unrelated objects which just happened to be on the same line of sight. For the variable star described here, this possibility may be ruled out due to the following reasons: 1) inspections of both the original CoRoT image (Fig. 5a) and the DSS image (Fig. 5b) reveal no other stars in the immediate vicinity of the variable and 2) pulsations disappear during the primary eclipse, which proves that the pulsating star is actually blocked from view.

3.2. Oscillating component classification

Following Moon (1986) we calculate reddening–free indices

$$[c_1] = c_1 - 0.19(b-y) = 0.777 \pm 0.087$$

$$[m_1] = m_1 + 0.33(b-y) = 0.155 \pm 0.111$$

which are consistent with a main sequence star of spectral class B8-9 or A7-F1 within the uncertainty (see Fig. 6).

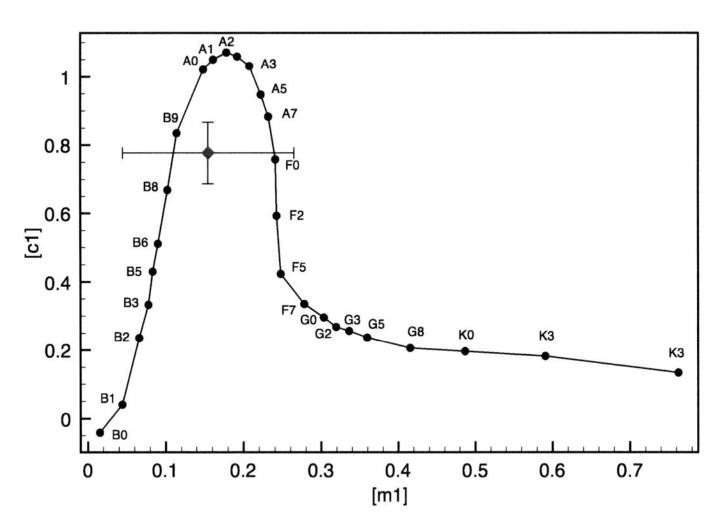

Figure 6: $[c_1]$—$[m_1]$ plot from Moon (1986) with the position of CoRoT 102980178 and its 1σ uncertainty indicated.

The Johnson B and V band observations with INT/WFC by Deleuil et al. (2006) (see Table. 1) have large error bars and where not used in the analysis, however, we note that the $(B - V)$ color of the star does not contradict our proposed spectral classification after accounting for the uncertainties in the Galactic extinction.

2MASS infrared colors can also be used to estimate the spectral class. Accounting for the uncertainty in the Galactic reddening discussed above, the infrared colors correspond to a spectral class in the range of early A – early M (Bessell & Brett 1988). However, it is possible that the secondary component of the binary system dominates the infrared light, so this spectral type estimation may correspond to the secondary star instead of the pulsating component.

The multicolor photometry excludes an early B spectral class, but is consistent with a reddened late B or F class star.

The period and amplitude of oscillations are typical for γ Doradus type but also within the possible range for young Slowly Pulsating B (SPB) and β Cephei stars. Information about the spectral class is crucial to distinguish between these possibilities: β Cephei variables are early–type B stars (B0 – B2.5), SPB stars have spectral types between B2 and B9 (Stankov & Handler 2005) while γ Doradus variables are typically early F type stars (e.g. Rodríguez 2002, Henry et al. 2007). A β Cephei classification can be excluded on the basis of multicolor photometry. Note also that according to Pigulski & Pojmański (2009), β Cephei variables are characterized by periods $< 0\overset{d}{.}35$ while the GCVS

(Samus et al. 2009) gives $0\overset{d}{.}6$ as the maximum period for this variability class. SPB classification is consistent with the Strömgren photometry, but SPB stars are usually characterized by longer pulsation periods of $0\overset{d}{.}5 - 5^d$ (Thoul 2009) and therefore, we consider the classification of the oscillating component as an SPB star to be less likely. γ Doradus stars are characterized by periods ranging from $0\overset{d}{.}4$ to 3^d (Kaye et al. 1999). Therefore, γ Doradus type variable is the most probable classification of the oscillating component.

3.3. Binary system modeling

For the construction of the binary system model we have cleaned the light curve from outlier points and corrected it for the linear systematic trend. No prewhitening for the pulsations was done. The oscillations should not affect the binary model because they are significantly smaller than the main features of the eclipsing light curve.

The light curve was analyzed with the *PHOEBE* software (Prša & Zwitter 2005, see also `http://phoebe.fiz.uni-lj.si`), which provides a user-friendly interface and additional capabilities to the Wilson & Devinney (1971) code. The software was recently modified to take into account CoRoT transmission functions, both for flux and limb darkening computations (for details see Maceroni et al. 2009). Since the orbit-unrelated variability is of much smaller amplitude and shorter period than the eclipses, we just fitted the light curve folded in phase according to the ephemeris presented above (Fig. 2).

After an initial screening of the parameter space, we proceeded to the light curve solution, using – at different stages – both the differential correction and the Nelder and Mead's Simplex algorithm. The fit adjustable parameters were: the inclination, i, the mass ratio, $q = M_2/M_1$, the surface potentials $\Omega_{1,2}$ (which, together with q, determine the components' fractional radii $r_{1,2}$), and the secondary effective temperature, $T_{\mathrm{eff},2}$. The albedo a_i and gravity darkening coefficients β_i were fixed at the theoretical values ($a_{1,2} = 1.0, 0.5$ and $\beta_{1,2} = 1.0, 0.32$, respectively, for stars with radiative or convective envelopes). We adopted a non-linear limb darkening law, whose coefficients were interpolated from the *PHOEBE* tables on-the-fly, according to the star's temperature and surface gravity ($\log g$). For the F-primary model (see below) we chose a logarithmic limb darkening law, for the B-primary one we chose a square-root law. The primary luminosity in the CoRoT passband, L_1, was computed at each iteration in *PHOEBE*, rather than adjusted, to enhance convergence (L_2 is derived from the other parameters and model atmospheres).

We started from a detached configuration but in all cases the iterative solution evolved towards a semidetached configuration (with the less massive star in contact). We checked, as well, the effect of non-zero eccentricity, which however did not provide any improvement of the fit.

It is well known that the solution of a single passband light curve is mainly sensitive to the $T_{\rm eff}$ relative values, therefore a shift in $T_{\rm eff,1}$ yields a shift in the same direction of $T_{\rm eff,2}$. It was therefore possible to fit the light curve incorporating different assumptions about the temperature of the primary (pulsating) component: $T_{\rm eff\ 1} = 7000\ K$ (early F spectral class) and $T_{\rm eff\ 1} = 11400\ K$ (late B class). As expected, we obtained different temperatures of the secondary: $T_{\rm eff\ 2} = 4692$ (K class) and $T_{\rm eff\ 2} = 6162$ (late F) for the first and second model, respectively. Apart from that, the models are very similar. The uncertainty of $T_{\rm eff\ 2}$ is dominated by our poor knowledge of $T_{\rm eff\ 1}$. For a particular assumed value of $T_{\rm eff\ 1}$ the formal error bars of $T_{\rm eff\ 2}$ calculated from the least square fit are less than 10 K.

To achieve an acceptable light curve solution the non-negligible O'Connell effect shall also be modeled, and for this purpose we introduced a bright spot on the primary component. Its parameters (location, size and temperature contrast factor, $T_{\rm spot\ 1}/T_{\rm eff\ 1}$) were not adjusted, but determined by trial and error, by putting the spot on the stellar equator and varying only the spot longitude and contrast factor (as the size and the latitude are highly correlated with them, see for example the discussion by Maceroni & van't Veer 1993). It is reasonable to expect the existence of such a spot as the result of mass accretion from the secondary component. However, different solutions (such as a dark spot on the colder secondary star, in this case due to surface activity), cannot be ruled out. The model parameters are summarized in Table 2.

Table 2: Model parameters of the binary system

Parameter	Model with F primary	Model with B primary
assumed $T_{\rm eff\ 1}$	$7000\ K$	$11400\ K$
$T_{\rm eff\ 2}$	$4692\ K$	$6162\ K$
$q = M_2/M_1$	0.206 ± 0.001	0.184 ± 0.001
i	$84\overset{\circ}{.}61 \pm 0.05$	$85\overset{\circ}{.}61 \pm 0.04$
r_1	0.1068 ± 0.0002	0.1064 ± 0.0002
r_2	0.2447 ± 0.0003	0.2363 ± 0.0003
L_2/L_1	0.7070 ± 0.0005	0.7183 ± 0.005
$T_{\rm spot\ 1}/T_{\rm eff\ 1}$	1.22	1.50

The spot angular radius was fixed to 15°, its colatitude to 90, the longitude of 240° and the temperature contrast factors in the table were found by trial and error. The uncertainties are formal fit errors (1σ).

One of the light curve features which are not reproduced by the models is a small difference in depth among primary eclipses. The eleven observed eclipses are not sufficient to draw a conclusion about a possible periodicity of the primary eclipse depth variation. The models also predict a symmetric light curve shape during the total eclipse phase which is not observed. These two effects may be caused by starspots on the secondary component. As can be seen from Fig. 2, there are some residual systematic deviations of the model with respect to observations. This will imply the presence of the orbital frequency (f_{orb}) and its multiples (typically the even ones) in the residual light curve spectrum which is discussed in the following section.

3.4. Pulsation frequency analysis

The residual light curve resulting from the subtraction of the binary model (version with early-F type primary) is presented in Fig. 3. It shows clear pulsation with a dominant frequency of ~ 2.75 c/d. If folded with the binary orbital period (see Fig. 4) the residual light curve shows a rather complicated pattern instead of uncorrelated data. This means that the main pulsation frequency is close to an integer multiple of the binary orbital frequency ($f1/f_{orb} = 13.898 \sim 14$). The pulsation disappears during the primary eclipse (phases $-0.05 - 0.05$ on Fig. 4), and this phase interval was excluded from the frequency analysis. In order to check for possible artifacts introduced due to omitting parts of the light curve, the analysis was repeated including observations during total eclipses. The introduced artifacts were found to be negligible.

The frequency analysis was performed using the *SigSpec* software (Reegen 2007), leading to 99 formally significant frequencies including the CoRoT orbit and its harmonics. The first few frequencies are listed in Table 3. The search for combination frequencies was performed using the software *COMBINE* developed by P. Reegen (see `http://www.SigSpec.org`). The frequency errors were estimated following Kallinger et al. (2008). The orbital frequency of the binary system was added to the frequency list, because from the plot of the resulting frequencies (Fig. 7) it is obvious that there are many combinations with the orbital frequency.

Only two frequencies (f1 and f3 in Table 3) appear to be genuine, which means not explained by a combination; f3 is close to f_{orb} but the frequency errors are small enough to conclude that the frequencies are significantly different (at 10σ level).

Analysis of the residual light curve after the subtraction of the B primary model leads to similar results (Table 3). The one major difference is that f8 becomes the second most powerful harmonic in the spectrum. Its amplitude is nine times larger compared to residuals from the binary model with an early-F type primary. f8 in the B primary model residual spectrum can be associated with the orbital frequency ($f_{orb} = 0.1978 \pm 0.0007$), the difference between them

Table 3: Frequencies detected in the residual light curve

	Model with F primary		Model with B primary	
	Frequency [c/d]	Amplitude [mmag]	Frequency [c/d]	Amplitude [mmag]
f1:	2.7494 ± 0.0008	9.52 ± 0.42	2.7494 ± 0.0009	9.68 ± 0.48
($4f_{orb}$) f2:	0.7917 ± 0.0008	7.48 ± 0.35	0.7920 ± 0.0009	6.31 ± 0.32
f3:	0.2115 ± 0.0012	3.28 ± 0.25	0.2181 ± 0.0021	1.97 ± 0.23
($2f_{orb}$) f4:	0.4004 ± 0.0015	3.19 ± 0.26	0.4012 ± 0.0016	2.57 ± 0.22
($10f_{orb}$) f5:	1.9784 ± 0.0017	3.27 ± 0.30	1.9783 ± 0.0014	3.86 ± 0.30
($8f_{orb}$) f6:	1.5823 ± 0.0017	2.81 ± 0.27	1.5822 ± 0.0013	4.17 ± 0.29
($13f_{orb}$) f7:	2.5657 ± 0.0019	2.48 ± 0.25	2.5650 ± 0.0019	2.32 ± 0.25
(f_{orb}) f8:	0.1978 ± 0.0031	1.00 ± 0.17	0.1940 ± 0.0008	9.05 ± 0.40
	...		...	

is less than 4σ. Most likely, it is not a real frequency but an artifact introduced by an imperfect binary model subtraction.

All the other frequencies can be explained by combinations mainly with the orbital frequency. This is what can be expected from a close binary because the primary component is likely to be distorted by tidal forces. If the F type primary hypothesis is accepted, the solution for the independent frequencies has two components: f1 and f3.

4. Conclusions

We have identified CoRoT 102980178 as a semidetached eclipsing binary with a $5\overset{d}{.}0548$ orbital period. It shows an Algol–type light curve with $1\overset{m}{.}26$ deep primary and $0\overset{m}{.}25$ deep secondary eclipses (in CoRoT photometric band).

Our light curve modeling leads to a semidetached configuration with the photometric mass ratio $q = 0.2$ and orbital inclination $i = 85°$. We note that the inverse problem of binary model fitting using only a broad passband light curve is degenerate and the proposed solution might not be unique.

The more massive component of the system is pulsating with the primary frequency of 2.7494 c/d ($0\overset{d}{.}36372$ period) and $0\overset{m}{.}01$ amplitude. The observed period and amplitude of pulsations as well as ground-based multicolor photometry favor its classification as a γ Doradus type variable, however, based on our data we cannot exclude the SPB-hypothesis for the primary. The detailed frequency analysis suggests the presence of an additional independent pulsation mode with the frequency of 0.21–0.22 c/d. Other detected frequencies can be explained as combinations of the above frequencies and the orbital frequency.

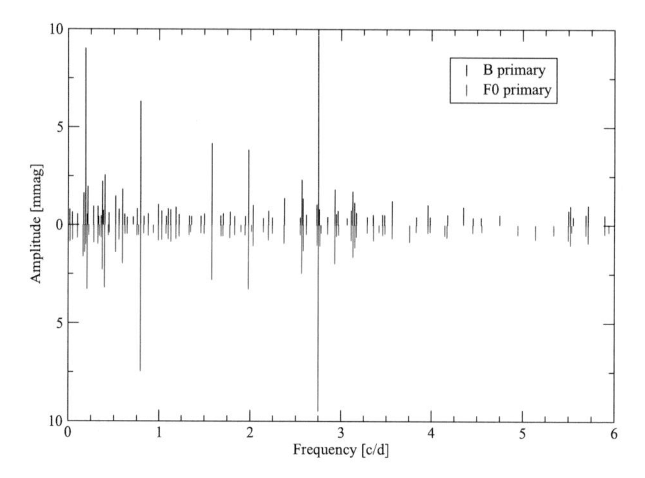

Figure 7: Amplitude spectrum of the residual light curve after subtracting models with B (black) and F (red, plotted downwards for better visibility) primary.

The system deserves a detailed spectroscopic study to better constrain physical parameters of the pulsating component. A Strömgren photometry of the primary eclipse could also be useful for this purpose.

Acknowledgments. Based on observations made with the WFC at the INT operated on the island of La Palma by the Isaac Newton Group in the Spanish Observatorio del Roque de los Muchachos of the Instituto de Astrofisica de Canarias. We would like to thank Sergei Antipin and Nikolai Samus for their aid in the classification of this peculiar object, Suzanne Aigrain and Werner Weiss for their help in establishing this fruitful collaboration, as well as Nicola Marchili, Frank Schinzel and the anonymous referee for reviewing this manuscript. KS was supported through a stipend from the International Max Planck Research School (IMPRS) for Astronomy and Astrophysics at the Universities of Bonn and Cologne. CM and CD research has been funded by the Italian Space Agency (ASI) under contract ASI/INAF I/015/07/00 in the frame of the ASI-ESS project. This research has made use of the Exo-Dat database, operated at LAM-OAMP, Marseille, France, on behalf of the CoRoT/Exoplanet program. The publication makes use of data products from the Two Micron All Sky Survey, which is a joint project of the UMass/IPAC-Caltech, funded by the NASA and the NSF, the Aladin interactive sky atlas, operated at CDS, Strasbourg, France, the International Variable Star Index (VSX) operated by the AAVSO and the NASA/IPAC Extragalactic Database (NED) which is operated by the

JPL, Caltech, under contract with the NASA. This research has made use of NASA's Astrophysics Data System.

References

Aigrain, S., Pont, F., Fressin, F., et al. 2009, A&A, 506, 425

Antipin, S. V., Sokolovsky, K. V., & Ignatieva, T. I. 2007, MNRAS, 379L, 60

Araujo-Betancor, S., Gänsicke, B. T., Hagen, H.-J., et al. 2005, A&A, 430, 629

Auvergne, M., Bodin, P., Boisnard, L., et al. 2009, A&A, 506, 411

Bessell, M. S., & Brett, J. M. 1988 PASP, 100, 1134

Bíró, I. B., & Nuspl, J. 2005, ASPC, 333, 221

Carpano, S., Cabrera, J., Alonso, R., et al., 2009, A&A, 506, 491

Chochol, D., & Pribulla, T. 2000, ASPC, 203, 125

Davidge, T. J., & Milone, E. F. 1984, ApJS, 55, 571

Damiani, C., Maceroni, C., Cardini, D., et al. 2010, Ap&SS, 66

Debosscher, J., Sarro, L. M., López, M., et al., 2009, A&A, 506, 519

Deleuil, M., Moutou, C., Deeg, H. J., et al. 2006, ESA Special Publication, 1306, 341

Fridlund, M., Baglin, A., Lochard, J., & Conroy, L. 2006, ESASP, 1306,

Handler, G., Balona, L. A., Shobbrook, R. R., et al. 2002, MNRAS, 333, 262

Hekker, S., Debosscher, J., Huber, D., et al. 2010, ApJ, 713, L187

Henry, G. W., Fekel, F. C., & Henry, S. M. 2007 AJ, 133, 1421

Ibanoğlu, C., Taş, G., Sipahi, E., & Evren, S. 2007, MNRAS, 376, 573

Kalberla, P. M. W., Burton, W. B., Hartmann, D., et al. 2005, A&A, 440, 775

Kallinger, T., Reegen, P., & Weiss, W. W. 2008, A&A, 481, 571

Kaye, A. B., Handler, G., Krisciunas, K., Poretti, E., & Zerbi, F. M. 1999, PASP, 111, 840

Khruslov, A. V. 2008, PZ, 28, 4

Kwee, K. K., & van Woerden, H. 1956, BAN, 12, 327

Lafler, J., & Kinman, T. D. 1965, ApJS, 11, 216

Liu, Q.-Y., & Yang, Y.-L. 2003, ChJAA, 3, 142

Maceroni, C., Montalbán, J., Michel, E., et al. 2009, A&A, 508, 1375

Maceroni, C. et al. 2010 in preparation; to appear in AN

Mkrtichian, D. E., Kusakin, A. V., Gamarova, A. Y., & Nazarenko, V. 2002, ASPC, 259, 96

Mkrtichian, D. E., Kim, S.-L., Rodríguez, E., et al. 2007, ASPC, 370, 194

Monet, D. G., Levine, S. E., Canzian, B., et al. 2003, AJ, 125, 984

Moon, T. 1986, Ap&SS, 122, 173
Pigulski, A., & Pojmański, G. 2009, AIPC, 1170, 351
Predehl, P., & Schmitt, J. H. M. M. 1995, A&A, 293, 889
Prša, A., & Zwitter, T. 2005, ApJ, 628, 426
Reegen, P. 2007, A&A, 467, 1353
Rodríguez, E. 2002 ESASP, 485, 331
Samus, N. N., Durlevich, O. V., et al. 2009, yCat, 1, 2025
Schlegel, D. J., Finkbeiner, D. P., & Davis, M. 1998, ApJ, 500, 525
Schwarzenberg-Czerny, A. 1991 MNRAS, 253, 198
Skrutskie, M. F., Cutri, R. M., Stiening, R., et al. 2006, AJ, 131, 1163
Stankov, A., & Handler, G. 2005, ApJS, 158, 193
Thoul, A. 2009, CoAst, 159, 35
Wilson, R. E., & Devinney, E. J. 1971, ApJ, 166, 605